How the Biosphere Works

How the Biosphere Works

Fresh Views Discovered While Growing Peppers

Fred Spier

CRC Press
Taylor & Francis Group
Boca Raton London New York

CRC Press is an imprint of the
Taylor & Francis Group, an **informa** business

First edition published 2022
by CRC Press
6000 Broken Sound Parkway NW, Suite 300, Boca Raton, FL 33487-2742

and by CRC Press
4 Park Square, Milton Park, Abingdon, Oxon, OX14 4RN

CRC Press is an imprint of Taylor & Francis Group, LLC

Library of Congress Cataloging-in-Publication Data
Names: Spier, Fred, 1952- author.
Title: How the biosphere works : fresh views discovered while growing
peppers / Fred Spier.
Description: First edition. | Boca Raton : CRC Press, 2022. | Includes
bibliographical references and index. | Summary: "This book offers a
simple and novel theoretical approach to understanding the history of
the biosphere, including humanity's place within it. It also helps to
clarify what the possibilities and limitations are for future action.
This is a subject of wide interest, because today we are facing a great
many environmental issues, many of which may appear unconnected. Yet all
these issues are part of our biosphere. For making plans for the future
and addressing our long-term survival and well-being, an integrated
knowledge of our biosphere and its history is therefore indispensable"— Provided by publisher.
Identifiers: LCCN 2021052892 (print) | LCCN 2021052893 (ebook) |
ISBN 9781032230412 (hardback) | ISBN 9781032230405 (paperback) |
ISBN 9781003275350 (ebook)
Subjects: LCSH: Biosphere. | Nature—Effect of human beings on.
Classification: LCC QH343.4.S684 2022 (print) | LCC QH343.4 (ebook) |
DDC 577—dc23/eng/20220126
LC record available at https://lccn.loc.gov/2021052892
LC ebook record available at https://lccn.loc.gov/2021052893

ISBN: 9781032230412 (hbk)
ISBN: 9781032230405 (pbk)
ISBN: 9781003275350 (ebk)

DOI: 10.1201/9781003275350

Typeset in Times
by codeMantra

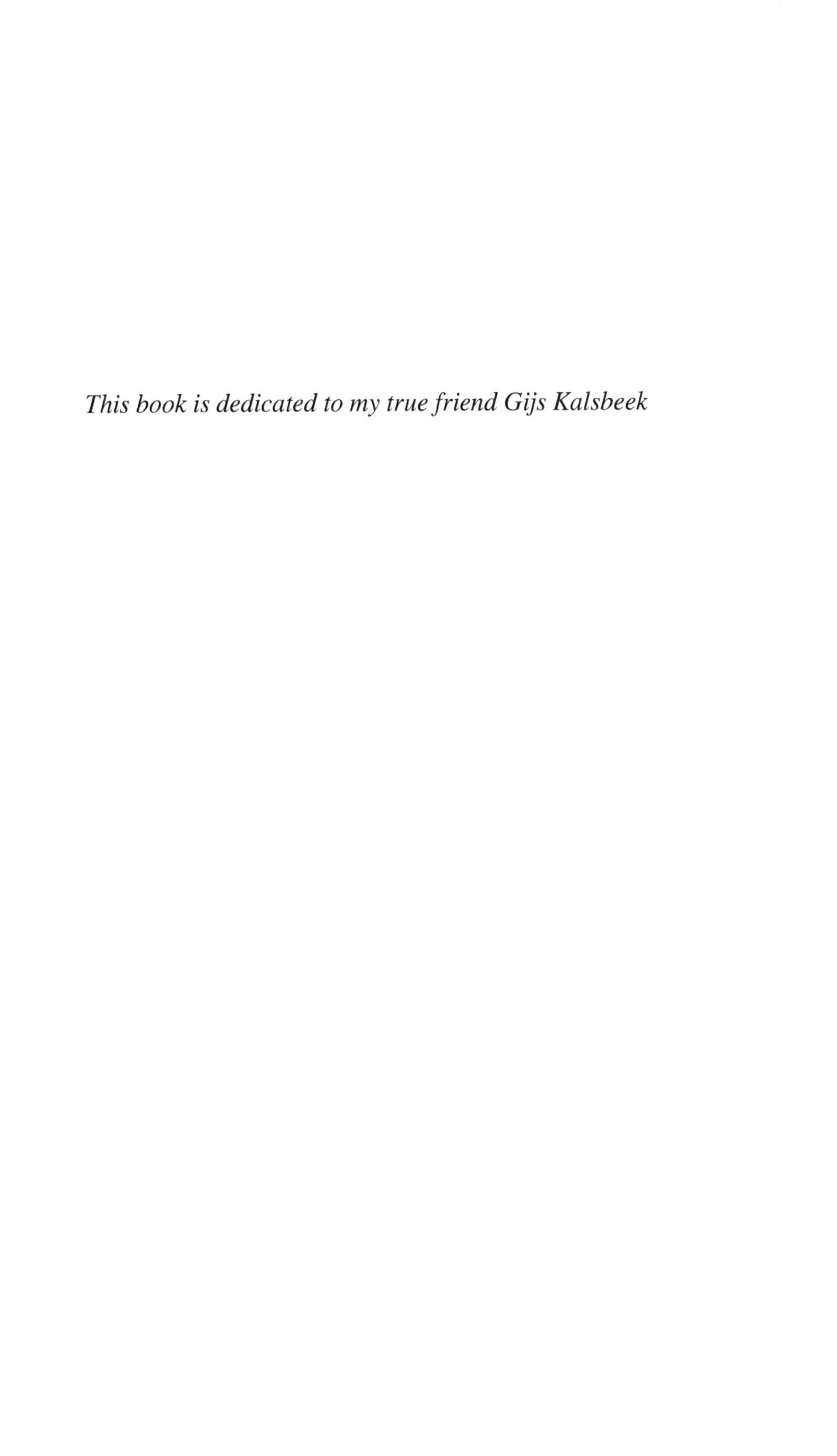

This book is dedicated to my true friend Gijs Kalsbeek

Contents

Preface

This book offers a fresh view of how our biosphere works, including its history. Those new ideas came as a direct result of growing pepper plants at home in Amsterdam, the Netherlands, in 2017, while observing what they were doing. That led to the discovery of simple general principles that appear to govern the biosphere and its history.

Such knowledge appears indispensable today. Human influences are increasingly affecting our biosphere, on which all of us as well as the future generations depend for their survival and well-being. Understanding our biosphere better, and, in consequence, helping to formulate forms of behavior that may assist us to preserve its favorable conditions, appears fundamental to me. That is my principal motivation to write this book.

I could not have made those observations and theoretical elaborations without a long personal history of interdisciplinary studies. Chapter 1 provides, therefore, my personal view of important pioneering scholars whose work is relevant to the theme of this book. Chapter 2 tells the story of how, by growing pepper plants, totally unexpectedly I found myself discovering those novel theoretical principles. Chapter 3 discusses theoretical concepts that are fundamental for understanding the biosphere's history from its very beginning all the way up until today. Chapter 4 mentions important theoretical issues that emerge while seeking to write such a history. Chapter 5 offers an account of the biosphere's history until the emergence of early humans. Chapter 6 discusses theoretical reflections on humanity's role within the biosphere's history. Chapter 7 tells the story of the biosphere's history during human history. Chapter 8 summarizes the current biospheric situation and discusses in general terms what could be done concerning the future. And finally, the Epilogue offers a few concluding reflections.

Because the book hopefully further explains itself, there is no need here for a more elaborate introduction. Instead, I would like to use this opportunity to thank all those friends and colleagues who have helped me over the past years to develop the ideas presented in this book.

First of all, my true friend Gijs Kalsbeek needs to be mentioned. Our friendship goes back to 1977, when both of us were finalizing our study of chemistry at Leyden University. Ever since that time, we have shared a great many aspects of life, including our ecological concerns and larger world experiences and the resulting changes of views, while engaging in countless and always stimulating conversations. As part of that, Gijs has been proofreading my work in progress for almost forty years, including earlier draft chapters of this book, while providing his always honest, detailed, and careful commentary. For all those reasons, this book is dedicated to him.

Also very important have been my friends and colleagues from Asturias, Spain, where they keep alive their wonderful tradition of wide-ranging, intelligent, sensitive, respectful, and often humorous conversations called *tertulias*. Those are first of all Olga García Moreno and Armando Menéndez Viso and their families, whose great hospitality, wonderful friendship, and *gran cariño* have very much helped me to feel at home in Asturias.

Other friends and colleagues from Asturias have contributed in similar ways. Mentioned here are Paz Pérez Encinas, Salvador Centeno Prieto, Manuel Gereduz Riera, Bruno Prieto and Laura Miranda González, Enedina García, Fabián Anuchnik Feldman, David Azpiazu Rodríguez, Diego Sanz Yus, Jaime Pérez González, Laudy Expósito Suárez, Yohanán Suárez Curieses, the Fundación Valdéz-Salas, as well as others not mentioned here by name.

Ever since the Dutch sociologist Joop Goudsblom and I started teaching our first big history course at the University of Amsterdam in December of 1994, I have greatly benefitted from the excellent expertise contributed by many lecturers, which has helped me to reach a better understanding of our biosphere's past as well as its place in cosmic history. They include the astronomer Ed van den Heuvel, whose most engaging lectures taught us a great deal about cosmic history, while his most loyal support for big history helped me to survive in academia. The astronomer Huib Henrichs' deep knowledge of the Solar System has also been enlightening.

The Earth scientist Harry Priem and the geophysiologist Peter Westbroek, both of whom are pioneers in the field of how to combine the history of Earth and life, also taught their wonderful lectures for a great many years. The always cheerful and engaging Belgian biologist Koen Martens contributed his excellent knowledge of biology, including the theory of natural selection as well as Charles Darwin's further work, which very much helped me to set sail on this current voyage. The expert geologist Jan Smit taught us a great deal about the evidence concerning the great extinction that ended the period of the dinosaurs as a consequence of a meteorite impact.

The paleontologist John de Vos, with whom we shared a great many beers during challenging discussions after his similarly challenging lectures, contributed not only his deep knowledge of the history of vertebrates—including humans—but also important general theoretical ideas mentioned in this book. Over many years, the climatologists Henry Hooghiemstra and Bas van Geel taught us a great deal about the history of the climate. The archeologists Jos Deeben and José Schreurs did the same in their field of early human history. Much of that knowledge was deepened during our always congenial conversations after the lectures.

Ever since she joined our little big history team in 2004, Esther Quaedackers has been an excellent and stimulating colleague, while she picked up slack during the years in which I had my hands more than full in taking care of our kids, for whom there was insufficient care available in the Netherlands.

In 1979 and 1980, my year-long stay among the hard-working, always friendly, and often humorous co-workers at the ecological farm of Gaiapolis located near Leyden introduced me hands-on to the possibilities and limitations of ecological farming. It was thanks through Gijs that I came to know them. And between 1985 and 1997, I owe even more to all the Andean Peruvian farmers who so kindly allowed me to share their lives, and, in doing so, provided me invaluable insights into their lives and perspectives. They include, most notably, my beloved *compadres* Julian Cconucuyca Florez and Paula Huallparimachi Mormontoy de Cconucuyca, and all our Peruvian family, as well as so many others in the village of Zurite, where I have not been for a long time, yet always hoping to return to as soon as the circumstances allow here. Also many other people in the Cusco area, most notably Dr. Jorge Villafuerte Recharte and Dra. Auristela Toledo de Villafuerte, have helped me a great deal to expand my world view while learning an awful lot about their liveways. Without them, this book would never have been written.

I feel gratitude to the great many expert academics whose work I read, all mentioned in the bibliography, while some of us met in person. Furthermore, my brother Hanz Spier as well as the biologist Isabel van Waveren contributed helpful suggestions. In Germany, Yvonne Fritz assisted me with her careful linguistic commentaries to draft chapters, while she also made me aware of important data, most notably the ecological passages in the German constitution.

The first public presentation of ideas elaborated in this book took place on May 14, 2019, at the University of Oviedo, Asturias, Spain, in Spanish, as the final lecture of the *Gran Historia* course organized by Olga García Moreno and Armando Menéndez Viso. It was titled *Nuevos desafíos en la gran historia: cuáles son y cómo afrontarlos* ('New challenges in big history: which ones, and how to deal with them'). The second public presentation was held on September 19, 2019, at the Birkbeck University of London, UK, in English, during the conference *Expanding World Views: Astrobiology, Big History, and Cosmic Perspectives* convened by the British astrophysicist and big historian Ian Crawford, titled: *Cultivating pepper plants: Fresh views on the history of the biosphere, life, and humanity.* I owe thanks to Ian Crawford, Michael Garrett, and Clément Vidal for their kind suggestions during that meeting.

I also owe special thanks to our daughter Giulia, without whom we would never have grown peppers plants. And last, but certainly not least, to my spouse Gina, for doing a magnificent job in providing our kids the best possible future in the United States.

Chuck Crumley and his expert team at the CRC Press kindly accepted my book proposal and turned it into the current book. For me, it is very rewarding that the CRC Press is publishing my latest work, which I see as the final result of a project that I started thinking of in the 1970s while studying chemistry, namely, to find out how humanity had ended up in its current ecological situation. At that time, I owned the 1971 edition of the *CRC Handbook of Chemistry and Physics*, also known as the 'Rubber Bible,' then and now, because of its authoritative position within the natural sciences. Little did I expect, not even in my wildest dreams, that fifty years later, the same press would publish my account of how the biosphere works.

As always, I am the only one who is responsible for this text. Any constructive comments for improvements will be greatly appreciated. I hope there will be many.

Fred Spier
Amsterdam, January 2022

Author

Fred Spier is senior lecturer in big history emeritus at the University of Amsterdam. He holds an M.Sc. in chemistry (with distinction) as well as an M.A. and a Ph.D. in cultural anthropology and social history (both *cum laude*, the highest distinction in the Netherlands). He has taught big history in various academic settings, while lecturing about it all around the world. He is the author of the highly acclaimed books *Religious Regimes in Peru* (1994), for which Spier was awarded a Praemium Erasmianum Study Prize 1993; *The Structure of Big History: From the Big Bang until Today* (1996); and *Big History and the Future of Humanity* (2010, 2015). Between 2011 and 2014, Spier was the founding vice president of the Intl. Big History Assn. (IBHA) and its president *de facto*, and between 2014 and 2016 its second president.

1 The Context of Discovery
A Personal Survey of the Natural and Social Sciences Concerning our Biosphere's History

The limits of man's knowledge in any subject possess a high interest, which is perhaps increased by its close neighborhood to the realms of imagination.

Charles Darwin, *Narrative of the Surveying Voyages of His Majesty's Ships Adventure and Beagle*, Vol. 3 (1839, p. 345)

It has been so easy to learn something of the Darwinian theory at second-hand, that few have cared to study it as expounded by its author. It thus happens that, while Darwin's name and fame are more widely known than of any other man of science, the real character and importance of the work he did are as widely misunderstood.

Alfred R. Wallace, 'The Debt of Science to Darwin.' *The Century Illustrated Monthly Magazine* (January 1883, p. 420)

INTRODUCTION

I do not know why I wanted to grow pepper plants in the spring of 2017. I just wanted to do it. Perhaps it happened because I had become accustomed to eating spicy food in Peru during my research visits there as a cultural anthropologist many years earlier. Peruvians of many walks of life love their pepper sauces, called *ajíes*, indispensable side dishes to almost any main course, breakfasts and desserts excepted.

During the spring of 2017 I was bound to the house because of the care that our daughter needed. That situation forced me to become creative with whatever initiatives were possible. Cultivating pepper plants might be such a plan, not least because our daughter loves pepper sauces. As a result, growing peppers might provide entertainment and education for both of us. As long as I remember, I had wanted to grow plants. Yet in the Netherlands, there had never been a chance to start my own vegetable garden.

So, in May 2017, while the Sun was becoming sufficiently strong, inside our Amsterdam apartment I sowed pepper seeds in a large white flowerpot, placed within what the housing corporation euphemistically calls the 'loggia,' a room with large windows facing southwest. The seeds came from red Spanish peppers; little yellow

DOI: 10.1201/9781003275350-1

Caribbean peppers; and green Mexican *jalapeños*, all bought at the Turkish grocery store around the corner. Their pods had been turned into *ajíes* using a traditional Peruvian recipe.

I had no idea what was going to happen, not even whether the seeds would sprout or not. I decided to let it happen while helping nature a little now and then. For instance, after having planted the seeds, the flowerpot was covered with thin transparent plastic foil to create a greenhouse effect. But other than that, I mostly wanted to see what was going to happen. Of course, it would be nice if the plants were to produce a next generation of peppers. But that was not the main goal. First of all, I wanted to observe what these plants were going to do.

Taking this more detached approach turned everything into an adventure. It did not really matter anymore whether the seeds were going to sprout or not, or how the plants would grow. Any result, even the most disappointing failure, would convey new information, from which we could learn, if interpreted correctly. And even if we could not find a good interpretation, the resulting confusion and realization of our own ignorance might also teach us good lessons. I had learned to adopt such an attitude while studying chemistry in the 1970s, experiencing many failures during my research.

I still remember the resulting frustration, for instance, in 1977 at the chemistry lab while trying to synthesize a substance called cyclical AMP that was going to be used for building artificial DNA molecules. While I was experiencing my umpteenth failure in making that molecule, my supervisor Jacques van Boom (1937–2004), an excellent chemist, saw my frustrations and told me: "What you need to do, is take that round-bottom flask and crash it into the sink. That will help you to get over it." I never did so, but his advice still remains etched in my mind.

Such initial confusion and realization of ignorance are inevitable when one sets out to experience, or investigate, unknown situations. That has happened to me many times in life, for instance, during my solitary backpacking travels overland across Europe, Africa, and India as well as during my academic research, first as a biochemist and later as a cultural anthropologist and social historian. In fact, I remember such feelings from much earlier in life, when I wanted to question so much that was taught at school. Yet back then, I never got a chance to even pose such questions, because that was not part of the Dutch school routine in those days.

It is understandable that teachers have limited time to answer questions from their pupils. But that may mean that our youngsters, often eager to learn, may be left with confusing thoughts and feelings. Let me give only one such an example. While going to primary school in the city of Eindhoven, situated in the Southern Netherlands, at around ten years of age, it struck me that virtually all the national history they taught us – even today still called 'patriotic history' – dealt, in fact, mostly with the more prominent Western Netherlands where Dutch independence had begun. But what had happened in the meantime in the Eindhoven area where we were living? Had nothing of any importance occurred there? And if so, who determined what was deemed important and what was not, and by which criteria?

All of that produced a great deal of confusion and awareness of ignorance inside of me, and it took me many decades to find satisfactory answers to those questions. In retrospect, teaching national history in such a way turned young people like me, who were living outside the supposed national core area, into outsiders. In doing so, a result was produced opposite to the one intended by national politicians, namely, shaping a shared national identity which helped all Dutch people to feel good about themselves and their past.[1]

Yet from such initial confusion and realization of ignorance, if questioned and pursued, fresh, and sometimes better, perceptions of reality may emerge. That requires a great deal of mental work. But almost invariably, such investigations produce positive effects, in this case my current personal summary of the city of Eindhoven's history, which can be found in Appendix 1 at the end of this book.

Pursuing my personal questions over a great many years has made me aware of the fact that most of us, including myself, live in a world, including its past, of which we know very little, even though we tend to navigate through life as if our daily surroundings are familiar. Doing so is part of our personal survival strategies. If we started questioning everything surrounding us, we would not be able to function very well anymore.

Such an awareness had struck me hard in May of 1986 after returning to the Netherlands from my first research visit to Andean Peru, where I had become familiar with local farmers' lives. Many of these people had become dear to me. Yet they lived their lives much like farmers had done in medieval Europe. It had been a major cultural challenge for me to become familiar with such a world.

But after I had become accustomed to their way of life, it actually seemed more 'normal' to me than my own life in the Netherlands, not least because those Andean farmers lived much closer to 'nature' than any urban people. In their lives, one experiences much more clearly where the food comes from, or how to build a house, because these people produce and make all of that themselves. In those closely knit communities, people are born, live, and die as part of local societies that are directly dependent on the surrounding natural environment.

As part of that, I remember very well a conversation in May 1986, near the end of my first research visit, with Paulino Tumpay, who had become a good friend of mine. I told him how scared I had been in the beginning, because I did not know what to expect, including whether I would be able to deal with Andean life. "Well, you see now," he replied, "everything is normal here."

I agreed. While, for instance, the women cooked their meals on very simple clay-baked stoves fired by locally-grown wood in what by Dutch standards would be considered very primitive kitchens, they were all very much people like you and me. Actually, it had often become extraordinarily pleasant to share life with them and be part of their daily routines, in which no one was excluded, while they generally

---

1 Only in December 2020 did I find on the internet that the armies of Prince William of Orange (1533-1584) –leader of the Northwestern Dutch uprising and still mentioned in Dutch school history as the 'Father of the Fatherland'– had committed considerable atrocities around the area where I grew up, most notably perhaps in the cities of Hertogenbosch and Roermond. Up until today, the authors of the national school curriculum may have considered it therefore better to avoid such controversies by not including that area's history at all during that particular period in time. Describing William of Orange as "Father of the Fatherland" was probably established in popular use by nineteenth and twentieth century Dutch historians seeking to reconstruct a proud history of the emerging Dutch nation-state.

treated me kindly. The rural Andeans would never leave me or any others alone, because that was considered socially undesirable.

How different was life after coming back to the Netherlands. Hardly anyone was interested in what I had experienced, while virtually everybody expected me to behave the 'normal' Dutch way. However, the Dutch urban way of life was no longer 'normal' for me, because I suddenly saw so much about Dutch life and its history that I had not seen before, while no one else around me appeared to be seeing any of that. As a result, it looked to me as if everybody in Holland whom I knew was acting like cultural robots without any awareness of the larger world. My good friend Tineke Luhrman, who shared some of those experiences in both Andean Peru and Ecuador, summarized this situation as follows: "If one has looked around in another people's dollhouse, one realizes that one lives in a dollhouse oneself, too."

So, suddenly there was a great deal of confusion and awareness of ignorance when, and where, I had expected it the least. Part of that included my realization of how little I actually knew about the Dutch ways of life and their history. My study of Andean history, while using a socio-scientific theory that worked very well, had taught me to look in novel ways that were rather different from any Dutch history that I knew. Yet my fresh ways of perceiving things were much more fruitful and insightful, or so I felt, at least for understanding the Andean ways of life and their history. But as a result, suddenly I did not understand my 'own' society anymore.

It has taken me decades to come to terms with all those thoughts and feelings, not least because Dutch social and academic life did not offer any room for such reflections, even among the Dutch cultural anthropologists whom I met at that time. As a result, I found myself very much locked into my own mental world. In fact, I am still working on integrating all those experiences into my personal life while investigating them further. Growing pepper plants in 2017 may well have been part of that effort.

To my delight, most of the seeds sprouted within a week and then started to grow. Because those tiny little plants were doing so well, very soon the plastic foil was no longer necessary. It was fascinating to watch what happened. The tiny sprouts expended their first efforts on growing a root to stabilize themselves in the soil and extract nutrients from it, while also racing to stretch out a little stem and unfold its first leave: its first solar panel. All of that was, of course, essential for survival. Without capturing solar energy, the little sprout would die very soon, as its seed contained only very little stored energy. And while roots do not extract energy from the soil, in addition to anchoring the plant they collect water and minerals that are needed for constructing and maintaining its complexity.

In doing all these things, a sprouting seed takes a considerable risk, because it expends all its stored energy and matter in the shortest possible time on turning itself into a little plant able to capture energy and matter. However, if the circumstances are not sufficiently right, for instance if there is not enough sunlight, water, or suitable soil, the young saplings will die very quickly. And because in nature the circumstances are often not sufficiently right, those plants need to produce a great many seeds to ensure their survival. If not, they will go extinct.

None of those observations is new, of course. But contemplating all of that made me realize that our pepper plants were doing what all of life has been doing during its entire existence on this planet, namely, seeking to survive. That is the principal goal of life. For pepper plants, producing seeds is not the main goal. It is simply a way of surviving the lean season when the plant itself is not sufficiently able to survive. Furthermore, making seeds promotes genetic diversity, which enhances their survival chances.

In other words, pepper plants are seeking to survive by using their own particular 'survival strategy' as it has evolved over long periods of time into its current form. To be sure, both a plant's 'survival strategy' and its evolution are not conscious processes. Such a survival strategy has become the way it is through the process of natural selection first outlined by Charles Robert Darwin (1809–1882) and Alfred Russel Wallace (1823–1913). In their theory, the struggle for life – so, the struggle for survival – plays a major role. Survival strategies can therefore be seen as the particular ways in which living beings wage the struggle for life. Curiously, to my knowledge, within biology this term has not yet systematically been used as a general analytical term.

Inspired by Darwin's famous book *On the Origin of Species* (1859), in 1864 the British academic Herbert Spencer (1820–1903) coined the term 'survival of the fittest,' which was later also adopted by Darwin. Yet within our big history course, the US biologist Frederick Schram (1943–) kept emphasizing that it would be better to speak of the 'survival of the fit enough.' That made perfect sense to me. For those who are not familiar with big history, this large-scale approach to history offers an overview of all of the past, from the beginning of the universe until life on Earth today.

By thinking along the lines of survival strategies, one suddenly sees all of life as consisting of a great number of species, which are all equipped with their own particular survival strategies, all of which have been honed into their present shape by the process of natural selection. This observation signaled the beginning of a growing number of fundamental questions that I began posing while watching our pepper plants grow. Yet to my surprise, to the best of my knowledge the theme of 'survival strategies' as part of the history of life has not yet systematically been explored.

Over the past 12,000 years, the survival strategies of plants and animals domesticated by humans have further been honed by human selection. That was no news to Darwin. In fact, the first chapter of *On the Origin of Species* describes the process of human selection of domesticated plants and animals as a model for explaining his novel concept of natural selection. Farmers such as my Peruvian friends and their ancestors have played an important role in that domestication process, most notably perhaps by domesticating potatoes, which today are feeding a considerable portion of the human world population.

As part of that domestication process, its emphasis changed. What matters for farmers is not solely the survival of their plants and animals, although that is also an important goal. Their main goal is the energy and matter that can be harvested from them. That is why humans began practicing agriculture and animal husbandry some 12,000 years ago.

As a result, I was suddenly looking at two different major goals: the major goal of all living organisms, which is simply survival, and the major goal of farmers for cultivating plants and keeping animals, which is the harvesting of energy and matter from them that is deemed useful for human survival and prosperity. Because farming has produced the currently-dominant forms of 'natural landscapes' encountered by urban citizens, a bias in perception may exist among such people that confuses those two goals. In other words, many urbanites may think that the 'goal' of pepper plants in the wild is to produce pepper pods and seeds, while in reality their major 'goal' is survival.

If we want to understand the history of life and humanity, we must be aware of such differences. This may well be common knowledge among biologists. Yet I do not remember having seen it clearly stated in such a way. This is the first bias, in this book called Bias # 1, that we will need to reflect upon as part of a growing number of biases that I encountered while growing pepper plants. In Appendix 2, a list of all these biases is provided,

There is more to it than that. As a big historian I had become aware of the huge theoretical issue that exists regarding the history of the biosphere and that of humanity. Basically, general theories uniting those two types of history do not yet exist. I had noticed that earlier, and had said so in two little movies that were recorded in 2016 as part of the University of Amsterdam's *Big History Massive Open Online Course* (MOOC) coordinated by my younger colleague Esther Quaedackers. This MOOC course can be taken on several internet platforms.

In 2016, I did not yet know how to resolve those theoretical problems and said so in those movies, while requesting support from the audience. Yet until today, I have not received even a single suggestion. However, by growing pepper plants and observing what they were doing I may have found a solution to this theoretical issue. Explaining that theory in more detail is the main subject of this book.

Before discussing that in the coming chapters, let us first consider what seminal inquiries into all the subjects involved in the biosphere's history and humanity's place within it had already produced, ranging from the biosphere itself, via physics and biology, all the way to the social sciences and human history. In doing so, let us begin by examining the idea of the 'biosphere.'

VERNADSKY'S BIOSPHERE

A major pioneer in studying the biosphere was the Russian-born scientist of Ukrainian descent Vladimir Ivanovich Vernadsky (1863–1945). Virtually unknown to the general public in Western Europe and North America, within the Russian Federation Vladimir Vernadsky has a similar status as Charles Darwin has worldwide. For instance, the 'International University of Nature, Society and Man,' known as such in 2005 when I visited it, is such a place. This university is located in the city of Dubna, situated to the north of Moscow. Since the 1940s, this city has been home to a major center for nuclear research.

When I was invited in 2005 by Russian colleagues to participate in a big history conference at that university, near its general entrance a lighted bust of Vernadsky was prominently displayed (of which I took pictures). That was not only lip service to that famous scholar. As part of that attention, all students were then required to consider the effects of all their projects on the biosphere, or so the dean told us in his formal welcoming speech. That encounter made me aware for the first time of the importance of Vladimir Vernadsky and his work.

Vladimir Vernadsky's bust as displayed in November of 2005 at the entrance of the International University of Nature, Society and Man, Dubna, Russian Federation. (Photo by the author)

What had Vernadsky done to receive such acclaim? In 1926, he had published his book *The Biosphere*, in Russian. In this study, Vernadsky elaborated the idea of the biosphere, the thin but very special layer covering Earth, situated between its lifeless inner layers and its cosmic environment. Vernadsky's emphasis was, most notably, on the great many changes effectuated by life on its geological surroundings, in doing so transforming a lifeless planetary outer shell into a biosphere.

Some of those perceptive ideas had already appeared in the little-known book *Hydrogéologie* (1802) written by the French naturalist Jean-Baptiste Lamarck (1744–1829). In that study, the French scholar used the word 'Biologie' for indicating the joint domain of all living beings, while it did not include inanimate matter.

The term 'biosphere' as elaborated by Vernadsky had already been coined in 1875 by the Austrian geologist Eduard Suess (1831–1914) at the end of his book *Die Entstehung der Alpen*, almost in passing on p. 159. Suess was a pioneering geologist, who came up with many novel concepts, including ideas that later led to the theory of plate tectonics. Yet the Austrian geologist never elaborated the concept of the biosphere in the ways of the great Russian scholar. In 1929, a French edition of Vernadsky's book was published, while in 1986, an abridged edition in English was published in the United States. Yet it was only in 1998 that a full edition in English saw the light.

By that time, the Cold War between the superpowers had ended, at least temporarily, thus facilitating more academic exchanges between the newly-emerging Russian

Federation and the rest of the world. Until that time, Vernadsky's seminal work had virtually been inaccessible within West European and North American scholarly circles, the French language area excepted. Vernadsky's only earlier publication in English about the biosphere was the short article 'The Biosphere and the Noösphere' in the journal *American Scientist* in January of 1945, republished in 1946 in *Main Currents In Modern Thought*. Those times were not favorable for widely disseminating such revolutionary ideas about Earth and its history.

Yet after in the late 1960s the first Apollo Moon flights had allowed humanity a view of Earth at a distance, the idea of the biosphere began to receive more attention in the USA. This happened, for instance, in the *Scientific American special issue* of September 1970 titled *The Biosphere*, which later that year was republished as a *Scientific American* book with the same title. In his opening article *The Biosphere*, the British-born US ecologist G. Evelyn Hutchinson (1903–1991) explicitly mentioned the contributions of Lamarck, Suess, and Vernadsky, in doing so paying them due honor. Yet this article did not contain a summary of all of Vernadsky's pioneering key ideas.

It was therefore only after the publication of the full English translation of Vernadsky's book in 1998 that his more detailed views became known within North Atlantic societies. Yet even today, his work is not widely read and valued outside of the Russian Federation and its former areas of influence. And sadly, the cover of the 1998 English edition is not as attractive as it could have been, which may not have facilitated spreading those innovative ideas.

In their analyses, both Suess and Vernadsky imagined themselves to be looking at Earth from space, which in their days was not yet possible. In fact, over the past two millennia, a range of eminent scientists and others have tried to imagine such a view, while contemplating Earth and its history, as outlined by the British historian Robert Poole in his book *Earthrise* (2008). Yet it was only in December of 1968, during the spaceflight of Apollo 8 to the moon, that humans for the first time saw Earth at a distance with their own eyes, while a few years earlier, unmanned spacecraft had already transmitted back to Earth such pictures. Those images stimulated a revolution in the ways people were thinking of – and feeling about – our home in space.

The importance of such distant views of Earth, imagined or otherwise, cannot be overstated. Any good academic analysis deals with details that are placed within a larger context. Doing so requires zooming in to see the details, and zooming out to see the context. It is not always necessary to zoom out all the way to see the entire universe. But if one seeks to clarify what a biosphere is as well as what makes a planet with a biosphere different from one without it, zooming in to see the details and zooming out to see Earth as part of the universe is absolutely required.

Let us return to Vernadsky's book *The Biosphere*. While it is impossible to do justice to any book's content in only a few lines, Vernadsky's major points as I understand them may be relatively easy to summarize. They are all related to the question of what distinguishes a planet with a biosphere from a celestial body that does not have one. In other words, what are the effects of life on planet Earth as a whole and vice versa?

Vernadsky's first answer was that a biosphere contains free energy accumulated by life. That may sound rather abstract, so let us try to explain what he meant by that. According to a well-known definition in science, free energy is the form of energy that is able to perform work. Food, for instance, contains free energy. That is the major reason why we eat it, because this free energy helps us to keep going. The energy that food contains was accumulated by plants, and also by animals that fed on the green stuff. This immediately explains why we cannot survive by eating only sand or clay, because those substances do not contain any useable free energy.

The concept of free energy was developed in the nineteenth century as part of the process of industrialization, which was driven by steam engines. At that time, a major issue was the question of how efficient such engines could be made, in other words: how much useable power could be produced by burning coal and making water boil. The search for answers produced a novel branch of science known as thermodynamics, the dynamics of heat. While further pursued, surprisingly thermodynamics turned out to be a rather fundamental science, because it was soon realized that those fresh insights were applicable not only to steam engines and internal combustion engines but also to the universe as a whole and everything in it. That was a major discovery.

While considering free energy, it is important to understand that the ability to perform work depends on the environment. For instance, sunlight radiating out from our central star does not contain any free energy on its solar surface, because it has the same temperature as its environment. In other words, on the Sun's surface there is no energy difference, no downward slope, along which this light energy can travel and perform work. But as soon as sunlight leaves the very hot solar surface and moves into cold and dark space, suddenly there is such an energy gradient. And the farther this solar radiation moves away from the Sun, the larger this gradient becomes, simply because in doing so space becomes colder and colder.

When this sunlight reaches Earth, the outside of which is not very hot – on average 15 degrees Celsius, about 59 degrees Fahrenheit – it contains a great deal of free energy that can perform work. This is the solar free energy that plants and microorganisms, including our pepper plants, capture to keep themselves going. And because sunlight is the most important source of free energy on our planet, those life-forms are the most important primary captors of free energy within the biosphere.

On Earth, there is another source of free energy, namely, geothermal energy, emanating from within our planet. And there are a great many life-forms, mostly microorganisms, that harvest this free energy. In doing so, those organisms are also primary captors of free energy. They practice today what is probably the oldest form of free energy harvesting on our planet. In fact, it is now thought that the first living organisms kept themselves going in such a way, while only later in time, life would also have learned to capture solar energy.

That was the point Vernadsky was making: all life-forms capture free energy and accumulate part of it within their bodies, stored within bio-molecules. Let us reflect on that for a moment. Not only all humans but also all other living organisms on land, in the water, and in the air contain free energy that was originally captured either from sunlight or from energy flows that are emanating from deep within our planet. A planet without life, by contrast, does not have such a reservoir of free energy accumulated by life.

Not solely living beings contain this captured free energy. All their leftover remains, including their excrements as well as whatever is left of their bodies after they die, may contain free energy. Within this context, we could think, for instance, of the coal layers and petroleum reservoirs that have intensively been exploited by humanity over the past few hundred years, because they contain so much concentrated free energy that can perform work for us. Doing so only works well when there is free oxygen in the atmosphere that can burn those substances. This free oxygen, that we breathe in with every inhalation to keep ourselves going, is a byproduct of capturing free energy from sunlight by plants and microorganisms. It is therefore also part of the free energy accumulated by life. That was also what our pepper plants started doing.

A second major point made by Vernadsky is that a biosphere contains atoms and molecules in places where they would not be expected without life. Atoms and molecules are the building blocks of which everything in the universe consists. Seen from a human perspective, these building blocks are very small. But they exist in enormous numbers. Every form of ordinary matter, including all of life, consists of atoms, while molecules are relatively stable combinations of atoms that tend to stick together.

In contrast to the inanimate universe, living organisms make complex molecules that are different from any form of matter that is occurring in the rest of the known cosmos. In other words, on a planet without life, none of its molecules would be arranged in such complicated ways. Because of the persistent actions of life, for instance, today Earth's atmosphere contains about 21% of free oxygen molecules and as little as 0.04% of carbon dioxide molecules. Without life, there would be no free oxygen in the atmosphere, while it would probably contain as much as 95%–96% of carbon dioxide. That is the current situation on our neighboring planets Venus and Mars. And there would not have been any plants and animals containing such unusual molecules, or coal and petroleum reserves.

Vernadsky's third major point was that life has been acting as a major geological force, which has been getting stronger during geological time as life grew and diversified. In his days, that was a revolutionary point to make, even though earlier naturalists, including the Prussian explorer and scientist Alexander von Humboldt (1769–1859) as well as Charles Darwin, had been aware of such influences.

Those earlier and more wide-ranging academic views in the first half of the nineteenth century were part and parcel of the fact that in those days, a clear academic distinction between biology and geology did not yet exist. It was only with the emerging specialization into various academic disciplines during the second half of the nineteenth century that biology and geology became separate fields of enquiry, to the extent that in the twentieth century, any mutual influences between life and geology were often no longer considered. As a result of this development, the earlier more comprehensive academic views were lost, while pioneers such as Suess and Vernadsky tried to restore them by using the novel concept of the biosphere.

One does not need a great deal of imagination to observe such influences, even within cities from which most of plant and animal life has been banished. Just take a look at a sidewalk, for instance, and observe what may be trying to grow there, including the efforts made by humans to prevent that from happening. If our species did not continuously engage in seeking to eradicate all those unwanted species, undomesticated life would soon overwhelm entire cities, if not the whole planet.

As the Dutch scientist Peter Westbroek (1937–) explained in his book *Life as a Geological Force* (1992), our planet's surface has deeply been influenced by life in a great many ways. This includes microorganisms and lichen, which almost everywhere can be seen 'eating' rock, by dissolving them with excreted acids. In doing so, they liberate atoms and molecules from those rocks that are useful to themselves, and, in consequence, also to the rest of living nature. But rocks do not contain free energy. As a result, lichen must obtain their free energy from other sources, usually solar free energy through the process of photosynthesis.

In the coming chapters, more will be said about Vernadsky's pioneering views. But for the time being, this summary may suffice. All the points made by Vernadsky could be observed while watching our pepper plants grow.

LOVELOCK'S BIOSPHERE

The British scientist James Lovelock (1919–) is another major pioneer in systematically thinking about the biosphere. As a NASA scientist in the 1960s, Lovelock made the suggestion that for detecting complex life on other planets it was not necessary to go there. One could instead study the light emitted by such a planet's atmosphere. If it were out of thermodynamical equilibrium – which means that its gases would contain free energy – then there must be life, because that is the only way free energy can continue to exist within an atmosphere. Without life, it would soon disappear, because it would bring about chemical reactions that moved the atmosphere toward a thermodynamical equilibrium. That would also happen to Earth's atmosphere as soon as life's existence ended.

To be sure, also without life, free energy flows, most notably sunlight and geothermal energy, push a planet's surface and atmosphere out of thermodynamic equilibrium. But life does so much more efficiently, because it actively captures such free energy and stores it within complex molecules. No other chemical process known to science does that similarly efficiently. As a result, a planet with life, especially one with more complex life, would have a biosphere with a composition of gases that are jointly out of thermodynamical equilibrium. By contrast, the planets Mars and Venus have atmospheres that are now virtually entirely in chemical thermodynamical equilibrium. As a result, those planets cannot possibly contain abundant complex life.

That was not a welcome message at a time when NASA was building its first spacecraft for exploring the Solar System. As a result, Lovelock soon had to find other ways for obtaining his free energy needed to make a living. Fortunately, he is a remarkably resourceful and talented scientist, which has helped him to come up with a series of inventions that have provided a sufficient income.

James Lovelock in 2005. (https://commons.wikimedia.org/wiki/File:James_Lovelock,_2005_(cropped).jpg.)

As part of those novel ideas, while still working at NASA, Lovelock began to develop his famous Gaia hypothesis, which states that all of life is not only interconnected, but that it also acts to regulate the biosphere in ways that are favorable to its own well-being. In doing so, the British scientist began to consider the whole planet as one single living super organism. Whether that is a useful concept or not remains to be seen.

Lovelock's pioneering insights were first published in his book *Gaia: A New Look at Life on Earth* (1979), which sported the full Earth photo on its cover taken in 1972 by the astronauts of Apollo 17, and subsequently also in *The Ages of Gaia: A Biography of Our Living Earth* (1988) and *Gaia: The Practical Science of Planetary Medicine* (1991). Especially in his more recent writings, Lovelock has been paying ample attention to humanity's influences within the biosphere, which, he thinks, will doom our species if it does not drastically change its behavior.

While developing his own views, Lovelock may not have been acquainted in-depth with Vernadsky's work. Yet the British scientist had clearly been thinking along similar lines, in both cases based on the science of thermodynamics.

CHARLES DARWIN AND ALFRED RUSSEL WALLACE'S THEORY OF NATURAL SELECTION

One of the biggest obstacles that held me back in thinking along the lines that will be explained in the coming chapters was the fact that the theory of natural selection formulated by Charles Darwin and Alfred Russel Wallace is often considered to be the central theory of the history of life. Yet the illustrious Darwin himself would never have said so.

To understand this, let us take a quick look at their theory of natural selection, as published in Darwin's famous book *On the Origin of Species* (1859). The central argument starts with the observation that all living populations always exhibit a certain degree of natural variation. In other words, no two individuals are exactly alike.

Much of this is genetically determined. In the second place, most, if not all species tend to produce more offspring than those which can survive, given the limited available resources as well as other circumstances that living organisms have to cope with.

In consequence, a process of 'natural selection' exists that favors those individuals which are sufficiently well-adapted to survive the struggle for existence, including producing a next generation with such favorable genetic characteristics. Charles Darwin called that 'descent with modification.' Directly following from this line of reasoning, Darwin concluded that all species would be related to each other in the form of one single family tree, which he called the 'tree of life,' within which all current species are descended from one single original species. Those revolutionary insights have helped a great deal to find order within the very complex history of life.

Charles Darwin, from: the *Allgemeine Illustrirte Zeitung* No 50, Stuttgart, 1871, p.1. (Copy in possession of the author)

While formulating his theory of natural selection, Darwin was inspired by the essay published anonymously in 1798 by the British economist and demographer Thomas Robert Malthus (1766–1834) titled 'An Essay on the Principle of Population as It Affects the Future Improvement of Society, with Remarks on the Speculations of Mr. Godwin, M. Condorcet, and Other Writers.' In this essay, which was later further elaborated, Malthus argued that because human populations tend to grow faster than the available food resources, over the course of time they will find themselves in great trouble, because they will not be able to feed themselves sufficiently well. As Darwin mentioned himself, his theory of natural selection was an application and further elaboration of Malthus's reflections on the future of human life.

The first formulation of Darwin's 'tree of life' was, in fact, an evolutionary interpretation of the work by the Swedish naturalist Carl Linnaeus (1707–1778), most notably his classification of all the known living species. During Darwin's life, a great many fossils had been recognized as extinct species that could not easily be accommodated within Linnaeus' scheme. While exploring parts of South America between 1831 and 1834 as a young adult, Darwin had also collected such fossils.

Furthermore, on that voyage around the world on the HMS Beagle between 1831 and 1836, the young British naturalist had seen firsthand how South American societies, and also the British, were marginalizing, if not totally eliminating, original populations in South America and the Pacific region. This was a brutal struggle for existence that was being won by the most powerful people of those times. Even though Charles Darwin was a rather peaceful and enlightened person who abhorred such violence, what he witnessed during that voyage in combination with the most recent geological, biological, and social insights inspired him to formulate his theory of natural selection.

While applying Darwin and Wallace's theory of natural selection to the enormous knowledge about the history of life gained since that time, it makes perfect academic sense today to think along those lines. That is why many biologists consider the theory of natural selection to be the central theory of the history of life. But is it such a theory? The British naturalist himself does not appear to have thought so. In his last book published in 1881 – more than twenty years after *On the Origin of Species* – Darwin focused on what he called *The Formation of Vegetable Mould, Through the Action of Worms, with Observations on Their Habits*. He saw the formation of vegetable mold as an emergent effect of life, as the unplanned result of the persistent actions of countless numbers of earthworms. In that book, the great naturalist did not even mention his theory of natural selection, even though he could have used it to explain some of his observations about earthworms.

Down House, Kent, UK, Charles Darwin's family home between 1842 and 1882, September 21, 2019. (Photo by the author)

A replica of the worm stone at Down House used by Darwin to measure soil changes resulting from earthworm activities, September 21, 2019. (Photo by the author)

In other words, the major reason for not considering the theory of natural selection to be the major theory for the history of life is that it does not explain such 'emergent effects.' Although this term did not yet exist in Darwin's days, the great naturalist clearly thought more broadly than only in terms of his theory of natural selection. As a result, apparently he did not regard it as the central theory of the history of life. Yet many of his followers started using it as such, perhaps mostly unconsciously. In doing so, they narrowed their perspectives on Darwin's work. Such a situation happens more often in academia, which is why it remains important to read the works of those pioneers, preferably in their original language.

I became aware of that theoretical bias while watching our pepper plants grow. How that happened will be explained in the next chapter. This is called the second bias in this book, so Bias # 2. To be sure, over the past decades several authors have recognized some of these limitations and have, in consequence, sought to expand the theory of natural selection. Such studies include Peter Corning's book *Holistic Darwinism* (2005), Susan Oyama's books *Evolution's Eye* (2000) and *The Ontogeny of Information* (2000), as well as the volume edited by her, Paul E. Griffiths, and Russell D. Gray titled *Cycles of Contingency* (2001). Yet to my knowledge, none of those authors has proposed the theoretical approach advocated in this book.

In Darwin and Wallace's theory of natural selection, no mention is made of the emerging science of thermodynamics, the application of which, in their days, was still mostly limited to purely physical processes.

A GENERAL THEORY OF THE BIOSPHERE'S HISTORY?

Ever since Darwin and Vernadsky, great progress has been made in describing and analyzing the history of our planet and life. Yet to my knowledge, even today no general theory has been proposed of the biosphere's history that would, in principle, structure and explain the history of life including its emergent effects, as an integral part of the biosphere's history. However pioneering and valuable Vernadsky's analyses are, they lack much historical depth, not least because at the time of writing, far less information as available.

By contrast, James Lovelock could base his Gaia theory on a much greater accumulation of such historical knowledge. As a result, his work contains much more historical depth. Yet however pioneering his Gaia theory is, it does not offer a general theory of the biosphere's history. A similar pioneering study *The Evolution of the Biosphere* (1986) based on Vernadsky's work by the Russian climatologist Mikhail Ivanovich Budyko (1920–2001) does not contain such a general theory either. And there are no other academic writings known to me about the biosphere's history that contain a general approach such as the one proposed in this book. An overview of the existing literature concerning this subject is provided at the beginning of Chapter 4.

As part of that situation, within studies of the biosphere and its history several academic fields often appear to be operating almost independently from each other, most notably those dealing with the evolution of species, of ecology, and of the biosphere as a whole. Take, for instance, the impressive tome *Fundamentals of Geobiology* (2012). It offers an enormous amount of detailed information on these themes. Yet it does not contain a general theory uniting all of that within one single theoretical framework, while it does not outline a history of the biosphere either. In its *Introduction*, Vernadsky's and Lovelock's contributions are mentioned. But their central ideas are not systematically explored, while Darwin's work is not mentioned at all.

As a result, there appears to be considerable theoretical confusion and ignorance concerning this rather important issue. In fact, today many academics may argue that searching for such a general theory would be futile, especially if one wanted to include the history of humanity, because the biosphere's history is so complex that one single theory structuring and explaining it will, in all likelihood, not be found.

Surely, before Darwin and Wallace, similar things could have been said – and were probably said – about the history of species. Yet those two scholars resolved that problem with a simple and elegant theory. In other words: a simple theory can help to structure and explain an – at first sight – bewildering diversity. Would it be possible to pull off a similar theoretical trick with the biosphere's history? That was the question that I posed in my University of Amsterdam Big History MOOC video of 2016, without having any answers at that time.

In retrospect, I did not yet have any answers at that time because, much like the great majority of scholars, I was put on the wrong foot by a number of academic biases, such as the idea that Darwin and Wallace's theory of natural selection was the central theory of the history of life. By watching our pepper plants grow, I became aware of this bias – and of ten more such biases –, all of which will be discussed in the coming chapters. By freeing myself from those biases step by step, I found myself on the track of formulating a general theory of the biosphere's history that is simple and elegant, at least in my opinion, not unlike the theory of natural selection formulated by our far more illustrious predecessors.

Before continuing with what happened to our pepper plants, we first need to take a quick look at theories of human history. That is necessary because we also want to consider the history of our species within the biosphere's history.

HUMAN HISTORY: MARVIN HARRIS AND CULTURAL ECOLOGY

While delving into cultural anthropology and human history in 1980, the author who influenced me the most was the US cultural anthropologist Marvin Harris (1927–2001). I read his textbook *Culture, People, Nature: An Introduction to General Anthropology* (1975) at night while working during the day on the ecological farm 'Gaiapolis,' situated near the Dutch city of Leyden where I then lived. I had received that book as a gift from a good friend of mine, Leony van der Splinter, who knew that I was considering to study cultural anthropology to find out more about human history and its current condition.

Marvin Harris's systematic treatment of culture greatly appealed to me, phrased as it was in terms of how humans deal with both the surrounding natural environment and their fellow human beings. The book took its readers all the way back to early human history, while the rest of human history was analyzed in terms of phases of social development, most notably those of early humans, followed by anatomically modern gatherers and hunters, and subsequently agricultural societies, tribes, state societies, and industrial societies.

By 1975, the age of computer informatization was only beginning to take off, and was therefore not yet included in Harris's analysis. It was this book that made me finally decide to enter the field of cultural anthropology, after a chance meeting in 1979 with the perceptive young German cultural anthropologist Joachim Theis on a train in the middle of the Sudan had put me on the track of that discipline.

Yet while in his book Marvin Harris provided a great many fascinating and challenging ideas about humanity's condition, he did not propose one single comprehensive theory of human history. I would have appreciated such a theory, after having obtained my M.Sc. in biochemistry at Leyden University. In the natural sciences such general theories – for example, the theory of the chemical bond – are extraordinarily important for structuring the perceptions and actions of scientists. Today, natural scientists would not know how to do science without such general theories.

While studying cultural anthropology at the Free University Amsterdam in the early 1980s, to my surprise none of those teachers particularly liked Harris's ideas. Most of them were instead focusing on their far more limited research interests, while espousing similarly limited theories. Like much of the social sciences and the humanities today, cultural anthropology was in those days an amalgam of small-scale theories without any systematic theoretical interconnections.

As a result, even today the social sciences and the humanities consist of a large number of competing theoretical schools, many of which appear to be socially mutually exclusive. Coming from the natural sciences, I was used to more unified theoretical approaches, within which smaller and more specialized theories ought to fit, while social exclusion as a result of adhering to one's particular theory does not appear to exist within the natural sciences as long as such theories are sufficiently supported by reliable empirical evidence. Therefore, I did not expect such a fractured theoretical and social situation within the social sciences, and understood it even less. Instead, I assumed that even with the inevitable specialization, all of us would aim to understand societies in a unified manner, especially after the Apollo photos of Earth at a distance had shown us such a unity.

The only cultural anthropologist at the Free University Amsterdam who, in those days, emphasized the importance of understanding local societies within their ecological context was Dr. Joop van Kessel (1934–). In his role as a Roman Catholic priest, van Kessel had gained his experiences of Andean societies in northern Chile. In 1980, he had defended his Ph.D. dissertation titled *Holocausto al Progreso: Los Aymarás de Tarapacá*, which presented a detailed investigation into the past and present of those people by carefully tracing the changes undergone by those societies over the course of their history. That was a most welcome approach. Because also Joop van Kessel's lectures resonated well, I decided to read his dissertation carefully, which became my first attempt at reading a book in Spanish.

As part of that interest, I became involved with van Kessel's projects in the Northern Chilean Andes, some of which I visited in early 1986 while I had to leave Peru to renew my visa. It was also Joop van Kessel who introduced me to his contacts in Cusco after I had taken the decision to do my fieldwork in the Peruvian Andes. Yet also he did not espouse more general theoretical approaches. His main interest was focused on the unequal distribution of power over the centuries between the native Americans and the dominant classes mostly of European origin as well as the resulting social and cultural effects.

It is not surprising that Joop van Kessel recognized the importance of ecology for understanding the history and present of those Andean people. In contrast to the rather flat, wet, and not very varied Western Netherlands, the South American Andes presents one of the most variegated, spectacular, and challenging, landscapes in the world. This has left very deep impressions on me, not least because the areas that I visited are all situated in the tropics, where the often strong and relentless sunlight, especially noticeable at the higher Andean elevations, very much accentuates these ecological and cultural differences.

In his lack of employing more general theories, Joop van Kessel was certainly not an exception. To the contrary, most social scientists and historians I know have tended to think similarly. Now and then, attempts are made within the humanities to develop overarching theories. Yet such efforts are looked down upon with great suspicion by the majority of scholars in those fields. Such a situation does not offer a great deal of encouragement for theoretical progress.

In the natural sciences, by contrast, the situation is very different. Yet that does not make a dent into the opinions of most practitioners of the humanities and the social sciences concerning a possible need for general theories. Usually, they argue that their fields are far too complex to find such theories. Yet in doing so, they do not appear to realize that a simple theory may yield explanations for a diverse and complex reality.

Surely, human history is the most complex aspect of the known cosmos. Yet also biological history is very complex. And, as mentioned above, a similar argument against a general theory underlying the origin of all species might have been raised before Darwin and Wallace came up with their simple and elegant theory. And the same argument is also applicable to, for instance, the modern theory of the chemical bond. The art consists of finding a simple yet efficient general theory that may help to structure and explain much, if not all, the diversity and complexity that can be observed in human history.

Fortunately, within the social sciences and the humanities a few exceptions exist, and Marvin Harris was a scholar who attempted to look much broader. I had the great fortune of meeting this fascinating man in Amsterdam on September 26, 1988, right before leaving for my second research visit to Andean Peru. This meeting took place at the Postdoctoral Institute of Sociology where I was then working on my Ph.D. thesis. One of my two supervisors, the Dutch sociologist Johan Goudsblom (1932–2020), had invited Marvin Harris to the institute to preside a symposium about his work, in which I participated. For many years, Goudsblom had been an admirer of Harris' work, which is sometimes controversial. Yet by sticking one's neck out theoretically now and then, others are challenged to think hard and formulate their positions. If one only follows the beaten track, not much theoretical progress will come as a result.

Marvin Harris and his wife Madeline at the Amsterdam Postdoctoral Institute of Sociology, September 26, 1988. (Photo by the author)

During that meeting, Marvin Harris was so kind to write encouraging words in my copy of his textbook, after I told him that it had made me decide to study cultural anthropology. That was very nice of him at a time when I was still at the beginning of my career in cultural anthropology. While doing such research, cultural anthropologists are almost totally dependent on the people they investigate. As a result, it is unpredictable whether one will come back with sufficient answers to one's research questions. In such a situation, the encouragement by a famous cultural anthropologist that I admired was very supportive.

Among most practitioners of the social sciences and the humanities, the science of thermodynamics as well as the insights by Darwin and Wallace, Vernadsky, and Lovelock, were, and are, mostly unknown. But even when they are known, they are usually considered virtually irrelevant for a better understanding of humanity's history as well as its present and future on this planet.

HUMAN HISTORY: NORBERT ELIAS AND HIS GENERAL THEORY OF SOCIETIES

In 1981, during my first year of studying cultural anthropology at the Free University Amsterdam, a Dutch introductory textbook on the social sciences put me on the track of Norbert Elias's sociology. That textbook carried the promising title *Mensbeelden en Maatschappijmodellen: kernproblematiek der sociale wetenschappen*, 'Images of Humans and Models of Society: fundamental issues in the social sciences.' On page 152, the author, the Dutch sociologist H. de Jager, presented a diagram in the form of a rhombus, on the sides of which all the (then) major socio-scientific theories had been positioned.

As an exception, the theoretical approach of one single author, de Jager explained, needed to be positioned right in the middle of that scheme, if it fitted it at all, because that theoretical approach was 'neither individual nor communal, neither materialistic nor idealistic, and neither micro- nor macro-sociological either.' That was the theory advanced by the German sociologist Norbert Elias (1897–1990). On the next page of that textbook, a few of Elias's salient points of view were summarized. Yet rather surprisingly, Norbert Elias's work was further hardly mentioned, while most of the attention was focused on the theories of more established social scientists, whose works had been positioned on the sides of that rhombus.

That curious situation immediately triggered my interest, because it made me feel that I might be on the track of something important. I decided to buy a few of Elias's books that had recently been translated into Dutch and started reading them. The first book that I studied carried the promising title *What is sociology?* In that book, Elias explained his general theory of interdependent people. The German sociologist perceived all people as connected through interdependency relations, which could also be understood as balances of power, which are changing over time in the form of processes.

These interdependencies are a general concept, and, as such, offer an interpretation of reality. They cannot be observed directly but can only be inferred through people's behavior. This situation bears great similarity to the concept of the chemical bond, which is also an interpretation of reality, that can be inferred through the behavior of atoms and molecules. This sociological model in terms of human interdependencies can be applied to all of humanity including its history.

As Elias explained in his book *What is Sociology?*, for emphasizing the importance of human interdependencies as a central theoretical principle he had actually been inspired by the idea of chemical bonds that keep molecules together. In other words, this was Norbert Elias' social version of that particular natural scientific theory, not unlike how Darwin had been inspired to apply Malthus' social ideas to biological nature.

Within this context, it may be important to know that before moving into sociology, Elias had first studied medicine. As a result, the German sociologist had a natural sciences' background, which he considered very helpful for understanding social issues. As part of this theory, Elias systematically placed humans within their surrounding natural environment, much like also Marvin Harris had been doing. Unfortunately, however, in contrast to Harris' outspoken theoretical position in this respect, Norbert Elias never elaborated that point very much.

This very short summary offers, of course, only a most cursory overview of Elias's sociology. Yet for our purpose it may be sufficient, at least in this chapter. During my Peru research, I employed Elias's general theory, together with a further elaboration that had been developed by the Dutch cultural anthropologist Mart Bax (1937–). I had met him in 1984 during a course on religion and politics that he was teaching at the Free University Amsterdam. Using Elias' and Bax's theoretical approaches worked very well during my Andean investigation, or so I felt, while my research experiences stimulated me to add a few more theoretical aspects that provided further clarity.

Thanks to the fact that, in 1987, also Johan Goudsblom had become interested in my work on Peru, the next year I obtained the Ph.D. position mentioned earlier, while Goudsblom became my second supervisor (together with Mart Bax). In the meantime, I had become aware of the fact that since the 1970s, Johan Goudsblom had been playing a prominent role in facilitating the publication and translation of Elias' work, in Dutch, English, and German. Furthermore, Goudsblom was constantly stimulating and assisting Norbert Elias personally in developing his ideas, while he had also offered him a place to live in Amsterdam, namely, upstairs in his own house. As part of those efforts, a circle of mostly young Dutch, British, German, and French sociologists interested in Elias's work had taken shape.

As a result, I suddenly found myself right in the middle of those social circles. Yet I kept distance from the old master himself, because I knew that he was becoming very old, and that he was in a hurry to write up everything that he could during the unknown but probably rather limited time that was left for him. I did not consider myself sufficiently important to request some of his precious time and attention, while I might not have to say enough that would interest him. Now, I regret that a little, but that is how I judged that situation in those days.

Norbert Elias delivering the final presentation at the Conference on Religious Regimes and State Formation, Free University Amsterdam, June 25, 1987. (Photo by the author)

Yet I met him once, on June 25, 1987, as part of the festivities celebrating his ninetieth birthday. We found ourselves walking together the short distance through the center of Amsterdam from Town Hall to the Palace on Dam Square, where another celebration was to take place. Even at his advanced age, Norbert Elias was still very bright and lively. After I told him a little about my Peru research at his request, he wanted to know why the indigenous people had not successfully risen against their colonizers.

That was a good question, which was not so easy to answer in only a few words, while keeping an eye on the rather unruly Amsterdam bicycle traffic. But his question kept me thinking for many years, and I still hear his voice posing that question ringing in my ears. I felt very impressed that, at ninety years of age, he would be that interested, lively, and inquisitive. Over time I have learned that many great scholars have tended to keep an open mind virtually all their lives. Charles Darwin offered such an example, and so did the great US world historian William H. McNeill (1917–2016), whose work will be discussed next.

HUMAN HISTORY: WILLIAM H. MCNEILL AND JOHN R. MCNEILL

Thanks to Johan Goudsblom, I became acquainted with world history as an academic approach, and, as part of that, with the work by William H. McNeill, most notably his big book *The Rise of the West: A History of the Human Community* (1963, 1991) as well as *Plagues and Peoples* (1976). Before that time, I did not know anything about world history as an academic discipline. None of my teachers had ever mentioned it, while, to the best of my knowledge, it was not part of any standard academic teaching in the Netherlands.

I met William McNeill for the first time on November, 20, 1992, during a symposium in Amsterdam. At that time, Goudsblom had just published his book *Fire and Civilization*. As part of its promotional activities, a symposium was organized. One of the keynote speakers was our US world historian, who, during the preceding years, had become a good friend of Goudsblom's, while I had been invited to react to McNeill's speech. Because I had no idea what he was going to say, I wrote him a letter requesting further information. All of this was happening before I had access to email. He sent me a kind letter back dated October 2, stating that "I have yet to read Prof. Goudsblom's book and until I have that privilege I can't tell what I have to say about it."

As a result, I was completely in the dark about what I was going to say during that symposium, while I had to present at least a few statements in the presence of those prominent scholars. Because I did not want to improvise right after McNeill's presentation, I prepared a short general statement which I delivered after he was done speaking. Fortunately, William McNeill appeared to like what I said.

By that time, I had successfully defended my Ph.D. thesis on Peru, on October 12, only five weeks before that symposium. During one of the friendly exchanges that followed, I gave McNeill a copy of my dissertation, expecting very little because he was probably busy with far more important things. So, I was totally stunned to receive a letter from him dated December 12 saying that "on the plane coming home I read your book and found it truly excellent; it most certainly should be published, and if you should find a letter to a publisher helpful I will be glad to provide one." Of course, I requested him such assistance, and subsequently received an incredibly supportive statement, part of which was reproduced on the book's back cover. It was subsequently published by Amsterdam University Press in 1994 thanks to his most generous support.

This was the beginning of a relationship between us that lasted until he very sadly passed away in July of 2016. We exchanged letters for more than twenty years, in which he first urged me to teach world history. And after Johan Goudsblom and I had begun teaching big history in 1994, William McNeill cheered us on, to use his own words, in his most generous inimitable ways. As he wrote me later, he wished that he himself had thought of such an idea, but that he had now become too old to engage in it.

Yet the great scholar did not react jealously at all, rather the opposite. Immediately he lent us again his unique support, most notably in 1996 after he had received the prestigious Dutch Erasmus Prize in Amsterdam. At the end of his acceptation speech, William McNeill announced that he donated half of the prize money to our initiative, an incredibly generous gesture! That produced a little wave of positive publicity for our project in the Netherlands.

William McNeill lecturing at the University of Amsterdam's big history course, December 11, 1996. (Photo by the author)

Over the years that followed we kept exchanging letters, in which he regularly urged me to write my own 'big book.' We also met personally whenever the circumstances allowed it. In 2013, at ninety-six years of age, he stopped writing because he felt he had become too old to do so. But I kept sending him postcards. Without his most generous support and encouragement, my career would not have developed as well as it did.

During William McNeill's regular visits to the Netherlands, Johan Goudsblom had introduced him to Norbert Elias and his sociological approach. That stimulated McNeill to rethink his own approach to world history and cast it into terms of interdependency networks, while paying ample attention to energy use. This led to the publication of a new book on world history, written together with his son John R. McNeill, who by that time had become an accomplished ecological historian. Published in 2003, this book carries the title *The Human Web: A Bird's-Eye View of World History*. This publication played an important part in the beginning convergence between sociology and world history.

By that time, John McNeill had published his influential study *Something New Under the Sun: An Environmental History of the Twentieth Century World* (2000), which was followed ten years later by *Mosquito Empires: Ecology and War in the Greater Caribbean, 1620–1914*. In that book, John investigated the influence of American infectious diseases, brought there by European colonialists, on the power relations between the Old and the New World, while adding fresh insights. This study built on his father's pioneering study *Plagues and Peoples* (1976), in which he examined the influence of diseases on human history. I have met John several times, and have great respect for him both as a person and as an ecological historian at Georgetown University.

William McNeill's grandparents were farmers, and, as a result, both he and John fostered an ecological awareness. William McNeill kept such a farmers' perspective all his life, as he explained in his charming little book *Summers Long Ago: On Grandfather's Farm and in Grandmother's Kitchen* (2009) written together with his daughter Ruth. His (unpublished) dissertation was about the history of potatoes in Ireland, a subject that remained one of his interests until the end of his life.

After his retirement from the University of Chicago, the McNeill family moved to Colebrook, Connecticut, where William McNeill began growing his own potatoes. He liked doing that in the afternoon, he explained to me, because doing that kind of physical work was a good way of reflecting on the writing that he had done in the morning. When I visited him in Colebrook in October of 2005 on his invitation for an overnight stay, he prepared us a meal consisting of potatoes and other vegetables that he had grown himself, while afterwards he showed me his vegetable garden. How much more moving world history can become than by experiencing all of that, I do not know.

HUMAN HISTORY: ERIC WOLF

Within the context of this book, another prominent US cultural anthropologist, Eric R. Wolf (1923–1999), deserves to be mentioned. Originally of Austrian descent, Eric Wolf had performed his fieldwork in rural villages in Mexico, Puerto Rico, and later also in the Tyrolean Alps of Italy, while placing his analyses within the larger framework of human relations. In his analyses, the great anthropologist combined his keen interest in cultural practices with the daily aspects of how to make a living.

In 1982, Eric Wolf published his groundbreaking book *Europe and the People without History*. This book offers an overview of human history focusing on what happened after Europeans began conquering the Americas and other world areas, paying attention to all the people involved, all of whom found themselves in interconnected processes. It was a reaction against the then dominant academic history in which true history began as soon as a society had produced its own written accounts, while all the rest was considered prehistory. Such an approach excluded from history all the societies without writing, including many of them conquered by Europeans.

After having studied rural societies in Mexico that consisted of mostly illiterate or semi-literate farmers, Eric Wolf realized the importance of farmers, herders, as well as gatherers and hunters for human history, even though such people had left comparatively few traces in the written record, and, as a result, had not received the historical attention that they deserved. By writing that book, Wolf wanted them to receive equal attention and respect in historical terms, while showing how much, and for how long, these rural people had been connected to dominant urban societies, to whom they had supplied a great many vital resources. This was part of his effort of seeking to write a more balanced human history, based on a theoretical approach in terms of power, dependency, and resource flows.

It was no coincidence that Eric Wolf wrote such a book. By studying rural villages in Mexico, and later by doing the same in his 'own' cultural backyard, Eric Wolf had learned to look at history from rural perspectives. Doing so offers rather different perspectives on history than the perspectives gained by studying urban life. While investigating rural Peruvian Andean life, a similar change of perspective had also happened to me. The cultural distance between the city center of Cusco and that of Amsterdam – a geographical distance of about 10,000 km – always felt considerably less to me than between the center of Cusco and Andean village life in Zurite, which are separated by only 35 km.

Furthermore, only during the fall of 2019, while investigating the city of Eindhoven's history summarized in Appendix 1, did I realize that my perception of my 'own' area of origin as a region where nothing of any importance had happened, offered another example of a 'people without history.' And by comparing the city of Eindhoven's history with Andean Peru's rural past, I realized that until the beginning of the industrial revolution those histories had actually been rather similar, especially with regard to how people had made a living. Seen from such a perspective, Eindhoven's regional history was, in fact, much more similar to that of Andean Peru than the history of the western part of the Netherlands. Had I been armed with such knowledge before starting my Peru research, the resulting cultural shock might have been considerably less.

In 1984, Mart Bax had recommended reading *Europe and the People without History* while teaching his course on religion and politics. During the preceding years, Mart Bax and Eric Wolf had become good friends. As part of that, Eric Wolf was invited to be a chairperson of the 1987 conference convened at the Free University Amsterdam where Bax's theory about religion and power was discussed. One can imagine my trepidation when I had to present my first anthropological paper on my Peru research within a session chaired by that famous cultural anthropologist. Fortunately, he treated me kindly, which was quite a relief, while during that conference we had a few further friendly exchanges.

As mentioned earlier, Bax's theoretical approach was based on Norbert Elias' work. The 1987 conference was dedicated to honoring the 90th birthday of that great sociologist. In his final speech, Norbert Elias did a wonderful job in presenting a paper, even though he could not read his notes anymore. So, he did everything from memory. I took photos and made an audio recording of that speech, both of which are still in my possession, now digitized. In fact, I had been appointed the official photographer of that conference. At that time, digital photography did not yet exist. That privileged position allowed me the opportunity of taking pictures of all those famous scholars.

Norbert Elias and Eric Wolf at the Conference on Religious Regimes and State Formation, Free University Amsterdam, June 25, 1987. (Photo by the author)

Afterwards, I sent Eric Wolf prints of photos that I had taken, most notably a picture of him together with Norbert Elias. They had already known each other for a long time, having met for the first time in an internment camp in England in 1940. That encounter first exposed Eric Wolf to Norbert Elias's thinking. There must have been some mutual influences over the following years, which contributes to explaining the development of Eric Wolf's ideas and actions.

The great US anthropologist responded kindly to my letter with the photos, which led to a few more exchanges over the years. He was also very supportive of my Peru research, while he later used my book *Religious Regimes in Peru* for his teaching at the City University of New York, as he informed me during a personal meeting in Amsterdam in January of 1997. He also reacted positively to my book *The Structure of Big History* (1996), as he wrote me on a postcard dated March 3, 1997.[2] That book came as a result of teaching big history at the University of Amsterdam starting in December of 1994.

[2] His comments on that postcard, after thanking me for sending him photos that I took of him in Amsterdam in January of 1997: "I am also still reading your Structure of Big History. It is very ambitious, but then why not be ambitious in tackling the problems that beset us. As I read on, the notion of "regime" becomes more familiar and manageable for me. I also wish you luck for your endeavors, and may the search for a global history prosper. My wife joins me in sending you all best regards and wishes. Yours, Eric R. Wolf" One wished that all historians and cultural anthropologists were equally supportive of such large-scale approaches to history.

COSMIC EVOLUTION: ERIC CHAISSON

While living in the charming little town of Laguna Beach, CA, in 1998, in the tiny apartment of my future wife Gina, I had bought a little laptop computer that provided us internet access through a regular telephone line. All of that was very new and exciting in those days. Thanks to the fact that Amazon.com – then still mostly an online bookstore – had begun adding second-hand booksellers to its website, I could search for such books using key search terms that were related to big history. That was how I found the books written by the US astrophysicist Eric Chaisson (1946–), most notably *Cosmic Dawn* (1981) and *The Life Era* (1987).

Until that time, I had thought that the Australian historian David Christian and I were among the first scholars teaching big history. But thanks to those discoveries I realized that someone else had preceded us (while subsequent internet research showed that there had been more early pioneers). I immediately ordered Chaisson's books and found that he had been a true pioneer of what he called *Cosmic Evolution*, which was very similar, if not the same, as 'big history.' Yet Eric Chaisson had preceded us by about two decades.

After reading Eric Chaisson's excellent books I succeeded in tracing his email address online. None of that would have been possible only a few years earlier. I wrote him an email, complimenting him with his books, and told him about our big history courses. After he kindly replied, I invited him to give a guest lecture in our Amsterdam course in the year 2000. He came indeed and did a wonderful presentation, still without PowerPoint, but instead using (now very old-fashioned) 35 mm slides, as all of us did at that time who used images to illustrate our lectures.

Eric Chaisson lecturing at the University of Amsterdam big history course, January 21, 2000. (Photo by the author)

During our further conversations in Amsterdam, Eric Chaisson mentioned the new book that he was writing about the rise of complexity in the universe, in which energy flows played an important role. He gave me a copy of the draft manuscript with the request to comment on it, which I subsequently did. That was another major challenge, because the intellectual level of that manuscript was extremely high. It was based on thermodynamics, yet without any knowledge of Vernadsky's work, of which I was not aware either at that time. Titled *Cosmic Evolution: The Rise of Complexity in Nature*, the book was published in 2001 by Harvard University Press. It is still a pioneering study in the field of cosmic evolution / big history, and it has helped me a great deal in formulating my subsequent theoretical views, including the ones explained in this book.

COSMIC EVOLUTION: HUBERT REEVES

I have never had the privilege of personally meeting the pioneering French-Canadian astrophysicist Hubert Reeves (1932–). I became familiar with his work thanks to the Dutch astronomer Ed van den Heuvel, who taught enlightening and engaging lectures in our big history courses for a great many years. When Johan Goudsblom and I started our first big history course in 1994, there was no big history literature available (or so we thought). So, we asked all the contributing lecturers to suggest required reading for the students.

Our astronomer Ed van den Heuvel suggested to use three chapters from the translation into Dutch of Hubert Reeves' book *Patience dans l'azur: L'evolution cosmique* (1981). In fact, Ed van den Heuvel had contributed a great deal to its translation and publication. That was an excellent suggestion, and we used those three chapters for many years. Of course, we had to ask Hubert Reeves for permission to do so. I succeeded in tracing his address in Paris, France, and sent him a letter with that request. On December 24, 1994, Reeves kindly sent back a handwritten fax document that I still have, saying that he was glad to give us permission to use the book in our course. That was a most generous gesture.

As a result of the enormous efforts required to keep big history going at the University of Amsterdam, for many years I did not have sufficient time to delve into Reeves's other studies. But when I finally did so, I found them truly enlightening. Especially his book *L'heure de s'enivrer: L'univers a-t-il un sens?* (1986), published in 1991 in English with the slightly more prosaic title *The Hour of Our Delight: Cosmic Evolution, Order, and Complexity* is a pioneering study on the rise of complexity within cosmic history, which has contributed a great deal to shaping my current thoughts.

ENERGY SPECIALIST VACLAV SMIL

Within the context of this chapter, also the Canadian energy scholar Vaclav Smil (1943–) deserves to be mentioned. In 2005, I had the privilege of meeting him after we had invited him to give guest lectures within our big history courses. As part of that, we shared a pleasant and enlightening conversation over a rather mediocre

Indian curry in Amsterdam. During his career, Vaclav Smil has written a great many illuminating energy studies, ranging from the biosphere to human societies. In doing so, he was building on energy studies by scholars from most notably the 1970s.

Vaclav Smil lecturing at the University of Amsterdam big history course, April 19, 2005. (Photo by the author)

Being of Czech descent, Smil was very much aware of Vernadsky's work. Yet he has not used it systematically, to the best of my knowledge, while I had then just begun my acquaintance with that great Russian scholar. For instance, while Smil paid considerable attention to Vernadsky and his work in his book *The Earth's Biosphere* (2002), this study is more a catalogue of a great many aspects of the biosphere than a coherent analysis of it written from a general theoretical perspective. That is similarly the case with all his other books that I have read.

Yet even though Smil's work may be lacking fresh general theoretical perspectives, it has provided a great many valuable data on the theme of energy, which has been very important for a better understanding of big history and the biosphere.

ENERGY SPECIALIST FRANK NIELE

Also the Dutch energy specialist Frank Niele deserves to be mentioned. In 2005, he published his grand overview of the history of life in the book *Energy: Engine of Evolution*. That is how I became acquainted with him and his work. We invited him in 2007, 2008, and 2009 to give guest lectures in our big history courses in Amsterdam and Eindhoven, which were well appreciated. At that time, Frank Niele was a principal scientist working for the Shell Oil Company.

Frank Niele lecturing at the Eindhoven University of Technology big history course, May 22, 2007. (Photo by the author)

In his book, Frank Niele did a wonderful job in presenting the case for a series of energy transitions during the history of life. Both Frank Niele's work and our conversations have considerably contributed to enhancing my understanding of energy. While writing his book, Niele was not aware of Vernadsky's work, while he mentioned the term 'biosphere' only once, in passing.

WORLD HISTORY: ALFRED CROSBY

Finally, but certainly not least, the work by the pioneering US world historian Alfred Crosby (1931–2018) deserves attention. Crosby's first interest was how the conquest of the Americas by Europeans had been possible, as well as how this had changed the world in both cultural and biological ways. Alfred Crosby summarized his findings in two books: *The Columbian Exchange: Biological and Cultural Consequences of 1492* (1972), and *Ecological Imperialism: The Biological Expansion of Europe, 900–1900* (1993).

In analyzing this subject, Crosby took the grand view, going all the way back to the last time in which all the landmasses had been joined into one single super-continent called Pangea. In doing so, he could build on the (then) fresh geological insights known as plate tectonics, the idea that over the course of time the landmasses had been moving across the face of the Earth. Later in life, Crosby examined the remarkable ways in which Europeans began to express ever more aspects of nature in numbers starting from the high Middle Ages, which was part and parcel of their growing power. This study resulted in his book *The Measure of Reality: Quantification and Western Society, 1250–1600* (1997). And after Crosby had become familiar with the translation into English of Vladimir Vernadsky's *The Biosphere*, he wrote his book *Children of the Sun: A History of Humanity's Unappeasable Appetite for Energy* (2006). The book title is a direct quotation from Vernadsky's book.

I met this great scholar only once, during a small conference on invitation in Hawaii in 2008. At that time, he was not in great health. Yet I was very much struck by his most intelligent, sensitive, and kind personality.

All of this, and much more, was in the back of my mind when I started growing peppers in the spring of 2017. What happened to our plants? And what triggered my current ideas? That will be discussed in the next chapter.

BIBLIOGRAPHY

Budyko, Mikhail I. 1986. *The Evolution of the Biosphere*. Dordrecht & Holland, D. Reidel Publishing Company (original edition in Russian 1984).

Chaisson, Eric J. 1981. *Cosmic Dawn: The Origins of Matter and Life*. New York, W. W. Norton & Co.

Chaisson, Eric J. 1987. *The Life Era: Cosmic Selection and Conscious Evolution*. New York, Atlantic Monthly Press.

Chaisson, Eric J. 2001. *Cosmic Evolution: The Rise of Complexity in Nature*. Cambridge, MA, Harvard University Press.

Chaisson, Eric J. 2003. 'A unifying concept for astrobiology.' *International Journal of Astrobiology* 2, 2 (91–101).

Chaisson, Eric J. 2004. 'Complexity: An energetics agenda.' *Complexity* 9, 3 (14–21).

Chaisson, Eric J. 2005. 'Follow the energy: The relevance of cosmic evolution for human history.' *Historically Speaking: Bulletin of the Historical Society* 6, 5 (26–28).

Chaisson, Eric J. 2006. *Epic of Evolution: Seven Ages of the Cosmos*. New York, Columbia University Press.

Christian, David. 1991. 'The case for "Big History".' *Journal of World History* 2, 2 (223–8).

Christian, David. 2004. *Maps of Time: An Introduction to Big History*. Berkeley & Los Angeles, CA, University of California Press.

Christian, David. 2011. 'The history of our world in 18 minutes.' *TED Session 6: Knowledge Revolution*, guest curated by Bill Gates: https://www.ted.com/talks/david_christian_the_history_of_our_world_in_18_minutes.

Corning, Peter A. 2005. *Holistic Darwinism: Synergy, Cybernetics, and the Bioeconomics of Evolution*. Chicago, IL, The University of Chicago Press.

Crosby, Alfred W. 1972. *The Columbian Exchange: Biological and Cultural Consequences of 1492*. Westport, CT, Greenwood Press.

Crosby, Alfred W. 1993. *Ecological Imperialism: The Biological Expansion of Europe, 900–1900*. Cambridge, Cambridge University Press.

Crosby, Alfred W. 1997. *The Measure of Reality: Quantification and Western Society, 1250–1600*. Cambridge, Cambridge University Press.

Crosby, Alfred W. 2006. *Children of the Sun: A History of Humanity's Unappeasable Appetite for Energy*. New York, W. W. Norton & Co.

Darwin, Charles. 1859. *On the Origin of Species by Means of Natural Selection, or the Preservation of Favoured Races in the Struggle for Life*. London, John Murray.

Darwin, Charles. 1881. *The Formation of Vegetable Mould, Through the Action of Worms, with Observations on Their Habits*. London, John Murray.

Darwin, Charles. 1958. *The Autobiography of Charles Darwin, 1809–1882. With original omissions restored, Edited with Appendix and Notes by his grand-daughter Nora Barlow*. London, St. James's Place, Collins.

Darwin, Charles. 2015. *Narrative of the Surveying Voyages of His Majesty's Ships Adventure and Beagle between the Years 1826 and 1836. Volume 3. Journal and Remarks 1832–1836.* Cambridge, Cambridge University Press (1839).

Elias, Norbert. 1978. *What is Sociology?* London, Hutchinson.

Goudsblom, Johan. 1992. *Fire and Civilization.* London, Allen Lane.

Harris, Marvin. 1975. *Culture, People, Nature: An Introduction to General Anthropology.* New York, Harper & Row.

Hutchinson, G. Evelyn. 1970. 'The biosphere.' *Scientific American*, 223, 3 (September issue), (44–53).

Kessel, Johannes Jacobus Mathieu Martinus van. 1980. *Holocausto al Progreso: Los Aymarás de Tarapacá.* Amsterdam, CEDLA Incidentele Publicaties, 16.

Knoll, Andrew H., Canfield, Donald E. & Konhauser, Kurt O. (eds.). 2012. *Fundamentals of Geobiology.* Oxford, Wiley-Blackwell.

Lamarck, Jean Baptiste. 1802. *Hydrogéologie, Ou. Recherches sur l'influence qu'ont les eaux sur la surface du globe terrestre; sur les causes de l'existence du bassin des mers, de son déplacement et de son transport successif sur les différens points de la surface de ce globe; enfin sur les changemens que les corps vivans exercent sur la nature et l'état de cette surface.* Paris, L'Auteur, Agasse, Maillard.

Lovelock, James E. 1987. *Gaia: A New Look at Life on Earth.* Oxford & New York, Oxford University Press (1979).

Lovelock, James. 1995. *The Ages of Gaia: A Biography of Our Living Earth.* New York, W. W. Norton & Company (1988).

Lovelock, James. 2000. *Gaia: The Practical Science of Planetary Medicine.* Oxford & New York, Oxford University Press (1991).

Lovelock, James E. 2006. *The Revenge of Gaia: Why the Earth is Fighting Back and How We Can Still Save Humanity.* London, Allen Lane.

Lovelock, James E. 2009. *The Vanishing Face of Gaia: A Final Warning.* London, Allen Lane.

Lovelock, James E. 2014. *A Rough Ride to the Future.* London, Allen Lane.

McNeill, John Robert. 2000. *Something New Under the Sun: An Environmental History of the Twentieth Century World.* London, Penguin Books.

McNeill, John Robert. 2010. *Mosquito Empires: Ecology and War in the Greater Caribbean, 1620–1914.* Cambridge, Cambridge University Press.

McNeill, John Robert & McNeill, William H. 2003. *The Human Web: A Bird's Eye View of World History.* New York, W. W. Norton & Co.

McNeill, William H. 1976. *Plagues and Peoples.* Garden City, NY, Anchor Press/ Doubleday.

McNeill, William H. 1986. *Mythistory and Other Essays.* Chicago & London, University of Chicago Press.

McNeill, William H. 1991. *The Rise of the West: A History of the Human Community; with a Retrospective Essay.* Chicago & London, University of Chicago Press (1963).

McNeill, William H. 1992. *The Global Condition: Conquerors, Catastrophes and Community.* Princeton, NJ, Princeton University Press.

McNeill, William H. 2005. *The Pursuit of Truth: A Historian's Memoir.* Lexington, KY, The University of Kentucky Press.

McNeill, William H. & McNeill, Ruth J. 2009. *Summers Long Ago: On grandfather's Farm and in Grandmother's Kitchen.* Great Barrington, MA, Berkshire Publishing.

Niele, Frank. 2005. *Energy: Engine of Evolution.* Amsterdam, Elsevier/Shell Global Solutions.

Oyama, Susan. 2000. *Evolution's Eye: A Systems View of the Biology-Culture Divide.* Durham & London, Duke University Press.

Oyama, Susan. 2000. *The Ontogeny of Information: Developmental Systems and Evolution.* Second Edition, Revised and Expanded. Durham & London, Duke University Press.

Oyama, Susan, Griffiths, Paul E., and Gray, Russell D. (eds.). 2001. *Cycles of Contingency: Developmental Systems and Evolution*. Cambridge, MA & London, England, A Bradford Book, The MIT Press.

Poole, Robert. 2008. *Earthrise: How Man First Saw the Earth*. New Haven & London, Yale University Press.

Quaedackers, Esther. 2016. *Big History MOOC*, University of Amsterdam. http://bighistory-platform.weebly.com/uva-mooc.html.

Reeves, Hubert. 1981. *Patience dans l'azur: L'evolution cosmique*. Paris, Éditions du Seuil.

Reeves, Hubert. 1986. *De Evolutie van het Heelal*. Amsterdam, Van Gennep.

Reeves, Hubert. 1986. *L'heure de s'enivrer: L'univers a-t-il un sens?* Paris, Éditions du Seuil.

Reeves, Hubert. 1991. *The Hour of Our Delight: Cosmic Evolution, Order, and Complexity*. New York, W. H. Freeman & Co.

Scientific American Staff (eds.). 1970. *The Biosphere*. San Francisco, CA, W.H. Freeman.

Smil, Vaclav. 1991. *General Energetics: Energy in the Biosphere and Civilization*. New York, John Wiley & Sons, Inc.

Smil, Vaclav. 1994. *Energy in World History*. Boulder, CO, Westview Press.

Smil, Vaclav. 1999. *Energies: An Illustrated Guide to the Biosphere and Civilization*. Cambridge, MA, MIT Press.

Smil, Vaclav. 2002. *The Earth's Biosphere: Evolution, Dynamics, and Change*. Cambridge, MA, MIT Press.

Spier, Fred. 1994. *Religious Regimes in Peru: Religion and State Development in a Long-Term Perspective and the Effects in the Andean Village of Zurite*. Amsterdam, Amsterdam University Press. A pdf of this book can be downloaded for free at: http://www.bighistory.info/bhi_005_035.htm.

Spier, Fred. 1995. *San Nicolás de Zurite: Religion and Daily Life of an Andean Village in a Changing World*. Amsterdam, VU University Press. A pdf of this book can be downloaded for free at: http://www.bighistory.info/bhi_005_035.htm.

Spier, Fred. 2010. *Big History and the Future of Humanity*. Oxford, Wiley-Blackwell.

Spier, Fred. 2015. *Big History and the Future of Humanity, Second Edition*. Oxford, Wiley-Blackwell.

Spier, Fred. 2019. 'On the social impact of the Apollo 8 Earthrise photo, or the lack of it?' *Journal of Big History* 3, 3 2019 (157–189). https://jbh.journals.villanova.edu/article/view/2425.

Suess, Eduard. 1875. *Die Entstehung der Alpen*. Vienna, Wilhelm Braumüller.

Vernadsky, Vladimir I. 1945. 'The Biosphere and the Noösphere.' *American Scientist* 33, 1, January issue (1–12). Republished in 1946: *Main Currents In Modern Thought*, April issue (49–53).

Vernadsky, Vladimir I. 1986. *The Biosphere* (abridged edition). Santa Fe, NM, Synergistic Press.

Vernadsky, Vladimir I. 1998. *The Biosphere*. New York, Copernicus, Springer Verlag. (1926, Original Version in Russian).

Wallace, Alfred R. 1883. 'The Debt of Science to Darwin.' *The Century Illustrated Monthly Magazine*, 25, January issue (420–433).

Westbroek, Peter. 1992. *Life as a Geological Force: Dynamics of the Earth*. New York & London, W.W. Norton & Company (1991).

Wolf, Eric R. 1966. *Peasants*, Prentice-Hall Foundations of Modern Anthropology Series. Englewood Cliffs, NJ, Prentice-Hall Inc.

Wolf, Eric R. 1982. *Europe and the People without History*. Berkeley, CA, University of California Press.

2 Growing Pepper Plants
Jointly Maximizing the Capture of Solar Energy

> The surface of leaves in forests and prairies is tens of times larger than the area of the ground they cover. The leaves in meadows in our latitudes are 22 to 38 times larger in area; those of a field of white lucerne are 85.5 times larger; of a beech forest, 7.5 times; and so on, even without considering the organic world that fills empty spaces rapidly with large-sized plants. In Russian forests, the trees are reinforced by herbaceous vegetation in the soil, by mosses and lichens which climb their trunks, and by green algae, which cover them even under unfavorable conditions. Only by great effort and energy can man achieve any degree of homogeneity in the cultivated areas of the Earth, where the green weeds are constantly shooting up.
>
> Vladimir Vernadsky, *The Biosphere* (1998, p. 78)

THE FIRST PHASE OF OUR PEPPER PLANTS' SURVIVAL STRATEGY: CAPTURING AS MUCH SOLAR ENERGY AS POSSIBLE

What happened next to our pepper plants? Within two weeks after sowing the seeds in early May of 2017, more than 90% of them had sprouted. This efficiency is remarkable. Apparently, pepper plants can produce complex things such as seeds in large numbers, and in such ways, that most of them work as soon as they find themselves in sufficiently good circumstances. This observation made me wonder how 'quality control' in nature works as well as how it might compare to the quality control in factories that produce large amounts of complex consumer items such as cell phones.

Every day, I carefully watched what the little plants were doing. Many of those observations have been known to science for a long time, yet they may be worth mentioning. The first thing the sprouts did was quickly develop both a stem and a root to anchor themselves in the soil while obtaining water and minerals through them. At the same time, they stretched and unfolded a tiny green leaf for capturing solar energy, while, unseen to our eyes, that little leaf also began to capture carbon dioxide from the air. All those capturing processes above and below ground level require large surfaces, which immediately explains why plants make a great many roots and leaves.

In fact, one can marvel at the plants' ability to suck carbon dioxide so efficiently out of the atmosphere, because it occurs there only in a very low concentration, about 0.04% of all the available air. That is not a lot. Sucking that up successfully must, therefore, imply large efforts at considerable energy costs. The concentration of carbon dioxide in the atmosphere is so low today (even though it has recently been

DOI: 10.1201/9781003275350-2

The pepper plants, May 21, 2017. (Photo by the author)

increasing as a result of human action), because organisms that engage in photosynthesis need to capture it to make their bio-molecules. For them, this odorless gas is their only available source of carbon. It cannot be extracted from the soil.

Within this context, it is important to mention that the chemical element carbon is a most important ingredient for constructing almost all the bio-molecules that life consists of. That is why all photosynthetic life needs to suck carbon dioxide out of the air. Over the course of our biosphere's history, their increasing numbers have led to a growing competition for this resource, which immediately explains why the concentration of carbon dioxide in the atmosphere is so low today. The concentration of that gas in the air is a joint emergent effect of life that cannot solely be explained by the theory of natural selection, although it comes as a result of it.

Let us return to our pepper plant observations. A seed contains its entire first plant structure: a root, a stem, and a first leave, all tightly packed together within it. Seeds also contain some stored energy to be used for the plant's first development. Why would plants make seeds? In this compacted way, plants can survive for a considerable time in adverse circumstances, most notably the cold or dry seasons, during which it is hard, if not impossible, to capture sufficient energy and matter as well as do useful things with them. The way seeds survive such difficult circumstances is to reduce their vulnerability and minimize their energy expenditure, while seeking to hang on until better times arrive.

The first small leave has a simple structure. It looks like a rather basic solar panel, not designed to capture a maximum of solar energy, but rather to function as well as possible in the largest possible range of circumstances. Apparently, they have evolved in such a way under pressure of the process of natural selection for surviving the vagaries of spring weather.

After having sprouted, there is no way back for the little plant. And without refueling, its energy and matter supply would soon run out, and the sprouts would quickly die as a result. For such plants, it is therefore imperative to get the time of sprouting right. That probably first of all depends on the ambient temperature and the humidity, which is why our little artificial greenhouse helped them in getting underway. In nature, the rising temperatures would probably also signal the return of solar energy that can be captured.

It is unknown to me how seeds detect all of that, biochemically speaking, as well as how they react to those signals in ways that make them develop into viable little plants. If a sprouting seed does not get its timing right – which may happen quite regularly in nature – it will wither away and die very quickly. In doing so, it will end a chain of life that started with the emergence of our first common ancestors, perhaps as long ago as four billion years.

In the process of sprouting, some difficulties related to its physiological makeup are encountered. For instance, while most pepper sprouts were successfully able to shed the seed's husk they came from, some were not. The least successful of them withered away quickly, because they could not successfully unfold their first solar panel, and thus rapidly ran out of their limited energy supply. Other sprouts were able to get partially rid of the husk, and were thus only partially able to unfold their first leaves. That meant obtaining less solar energy at that rather crucial period in their early development. As a result, those less fortunate little sprouts spent more precious time in getting themselves settled into the soil while using whatever limited solar energy they could obtain.

In consequence, compared to the more successful sprouts, those strugglers found themselves at an immediate disadvantage in the race to obtain as much solar energy as possible for further growth and prosperity. Yet some of them made it anyway, as we observed during their further growth. In sum: only those individual saplings survive and prosper that are sufficiently equipped to function within the prevailing circumstances. In nature, all of this has been honed by the process of natural selection, and, in the case of our pepper plants, more recently also by the process of human selection through domestication.

What would be the efficiency of seeds sprouting successfully in situations that are not controlled by humans? How many seeds would sprout and turn into full-blown pepper plants in nature, and how many of them would not make it? This probably varies according to the circumstances. In addition to the inevitable variation in weather from year to year, one can imagine, for instance, that some seeds were spread by birds which ate the peppers and that, as a result, those hardy travelers quickly reached areas that may have been either less or perhaps more favorable, depending on the circumstances.

In such a process, the seeds might spread fast and wide as long as there were enough winged carriers that would take them somewhere else. That is why, I suspect, the pepper pods turned brightly colored after the seeds had sufficiently ripened, so as to attract their flying friends. In more recent times, humans have taken on the role of transporting them to other places, which is how they ended up in our flowerpot, while birds were kept at a distance.

After the young and very active little pepper plants had done all these things, they began to grow rapidly, beyond our wildest expectations. By quickly making more and larger leaves, they could capture increasing amounts of solar energy. This led to a positive feedback loop, which put those plants on a nonlinear growth track. Doing so is very helpful in situations in which other plants may behave similarly. In other words, such a nonlinear growth survival strategy may come as a result of the competition for the available solar energy within and among species, in doing so again producing a process of natural selection.

HOW EFFICIENT ARE LEAVES?

Seen from an engineering point of view, pepper plant leaves are very efficient structures. Constructed with a backbone mostly consisting of carbon, hydrogen and oxygen, their shapes are maintained by using water under pressure. Much like in almost all organism including ourselves, water forms the largest portion of the leaves. As soon as it is lacking, the leaves will shrivel away. And in contrast to human-made solar panels, plant leaves are perfectly recyclable.

In 2017, we did not investigate this any further. But on August 19, 2020, while again growing pepper plants, this time on my own, I decided to measure this by cutting off a few large *Capsicum annuum* leaves and weighing them. In trying to do so, it turned out that none of my scales was sufficiently sensitive to accurately weigh one single leave, because it was too light. I therefore cut off ten leaves and jointly weighed them, which yielded a combined weight of about 14 g.

I next measured their combined surfaces by spreading the leaves out and making them fit together as well as possible. This yielded an approximate circle with a diameter of about 30 cm. Their joint surface was, therefore, about 700 cm^2. Because their combined mass was about 14 g, their solar-capturing surface divided by its mass was about 50 cm^2/g.

How would that compare to human-made solar panels? In 2020, according to the website www.energysage.com, the average solar panel has a surface of about 65 by 39 inches and a mass of some 40 pounds. The solar-capturing surface per gram of such panels is therefore 0.9 cm^2/g, while, as we just saw, that of pepper plant leaves is about 50 cm^2/g. Apparently, pepper plant leaves are about fifty-five times as efficient as human-made solar panels in terms of the needed mass per surface area.

As mentioned, those leaves mostly consist of water. And as long as water is sufficiently available, it is a cheap construction material. How much water would those leaves contain? To find out, the same leaves were dried between the pages of a notebook. About one month later, their combined mass was only about 3 g, which means that they had contained 79% water. This dry weight yielded a surface per gram ratio of 233 cm^2/g. So, by only looking at their dry mass, the surface/mass ratio of *Capsicum annuum* pepper plant leaves would be about 260 times larger than that of human-made solar panels.

Yet in contrast to those human-made constructions, which do not require energy to stay in the same shape, plants do have to expend energy to pump up the water and keep the leaves pressurized, while they also need water for cooling the leaves through evaporation, especially on hot and sunny days. I do not know how much energy that would cost, but obviously, such a requirement adds to the plant leaves' construction and maintenance costs, while this can only be done well in situations in which both energy and water are sufficiently abundant.

What would be the efficiency of leaves in capturing solar energy? According to data found on the internet, a typical efficiency of plant leaves doing so would be about 5%, while, in 2020, average solar panels would capture as much as 23% of the available solar energy. In that respect, human-made solar panels are about five times as efficient as plant leaves. Yet by factoring in the mass of the construction materials excepting water, plant leaves would still be about forty times as efficient as human-made solar panels. So, in this respect, our pepper plant leaves clearly outperformed any current human engineering.

It may take a while before human cultural achievements will catch up in that respect with what natural selection has produced over billions of years. Apparently, it costs pepper plants relatively little matter and energy to make leaves while the yields in terms of capturing solar energy are large. This immediately explains how pepper plants can make so many leaves so quickly as soon as the circumstances are sufficiently right.

NONLINEAR GROWTH OF OUR PEPPER PLANTS

Let us return to our pepper plants in 2017. While they began growing rapidly in May, the sun was getting higher in the Amsterdam sky every day, while daylight time was similarly increasing. This meant that every day, increasing amounts of solar energy became available to our pepper plants, which stepped up the speed of their nonlinear growth. Seen from a global point of view, the fastest and largest changes of the amounts of sunlight reaching any place on our planet's surface occur near the spring and autumn equinoxes. This comes as a result of Earth's orbit around the Sun and the angle of its axis. So, the cosmic parameters of Earth as it swings around the Sun determine to a large extent the amount of solar energy that our pepper plants could harvest.

This rapidly increasing sunlight during the spring also led to an increase of the ambient temperatures. This speeded up the biochemical reactions of our pepper plants. The increasing velocity of chemical processes caused by higher ambient temperatures is also a nonlinear process. While studying chemistry, I was taught as a rule of thumb that, on average, an increase of 10°C doubles the velocity of chemical reactions.

For instance, in the Peruvian Andean city of Cusco, situated at about 3,300 m (10,000 feet) above sea level, it takes about twice as long to boil potatoes as it does at sea level. Boiling potatoes means that they are undergoing chemical reactions. At the altitude of that ancient Inca capital, the air is considerably thinner, which causes a lower boiling point of water. As a result, in Cusco, water boils already at 90°C, instead of 100°C at sea level. This doubles the time needed to boil potatoes in Cusco. So, how long would it take to boil potatoes on top of Mount Everest? I will leave that calculation as a little challenge to the readers.

While observing our pepper plants grow, it struck me that certain favorable circumstances –in our case the increasing supply of solar energy in spring – can keep improving even though that energy is also being captured at increasing rates. This can happen thanks to the fact that the Sun's energy output is not dependent on any consumption of it far away. But it certainly helped that the Amsterdam spring and summer of 2017 offered unusually sunny weather conditions. Because the windows of our apartment are oriented toward the southwest, the plants would only receive direct sunlight for less than half a day at most. But that was enough for the little plants to grow well. Starting on May 21, 2017, I began to take photos of their development. In retrospect, I should have started right at the beginning, which I did in 2020 when I conducted a similar experiment.

Very soon, the plants became taller while growing more and larger leaves. By the beginning of June, I decided to support them using skewers. In natural circumstances that might not happen, even though I do not know this for sure, because young pepper plants might find other plants to support them. Yet in our little apartment, I did not want those plants to be hanging over the edge of the flowerpot and, in doing so, potentially produce a mess in the loggia. Supporting their pepper plants in such ways is what horticulturists do who grow them. This made me wonder how undomesticated pepper plants would grow in their natural environment. Would they stay near the ground, or would they use the support of other plants wherever and whenever possible?

The plants needed to be watered almost every night. Without doing so, the leaves began to wither soon. Apparently, the increasing surface of all the leaves combined did not only lead to a growing capacity for capturing sunlight and carbon dioxide, but it also led to a growing evaporation of water. Apparently, those delicate (and thus relatively 'cheap') leaves had not developed any protection schemes to counter this. Clearly, those pepper plants had evolved to grow in circumstances in which there was not only enough sunlight but also sufficient water.

No insects or other organisms settled on the plants to extract energy and matter from them for supporting their own survival. This was probably first of all the result of a lack of such organisms within our apartment. Grown outside, pepper plants do suffer from invasions by numerous organisms that first of all attack their leaves, which are by far the largest and most vulnerable portions of those plants, where, in consequence, the circumstances for harvesting of energy and matter are good for the predators.

Attempts to offset such potential losses may be another reason why pepper plants produce so many leaves. In other words, their survival strategy may have evolved into the direction of producing more and cheaper leaves rather than investing in sturdier and more expensive solar panels, not unlike how humans may opt for cheaper clothes that wear out more quickly rather than buying more durable, yet often more expensive, brands.

But quite possibly, pepper plants may also have evolved ways to defend themselves by producing certain chemicals in their leaves that are toxic to their predators. In fact, most plant leaves cannot be eaten raw by humans for exactly such a reason. That would probably also include the leaves of pepper plants. We did not try to eat them, out of respect for such possible poisons, and I am not aware of any societies that do consume them. This makes one wonder how domesticated plants such as lettuce and spinach evolved, the leaves of which are eaten by humans. Whatever the case, the possible production of toxins by pepper plants to defend themselves would imply an investment in complexity that must have an energy and matter cost attached to it.

In their continuously-evolving survival strategy, pepper plants are balancing costs and benefits, much like all of life does. All of that is again honed by the process of natural selection. As part of that, those plants find themselves on a dependency path from which they cannot easily escape. Like all other living organisms, there is only one way to go, namely building on what already exists. For complex living beings such as pepper plants and humans, downsizing their complexity is usually very difficult, if not impossible, unless they find themselves in special circumstances that allow, or even favor, such a change.

Today, such rather special circumstances are, for instance, created in biochemistry laboratories. Between 1975 and 1977, during my biochemistry study I participated in such a project that sought to grow tobacco plants from single leave cells that had been stripped of their cell walls. Those naked single cells were called protoplasts, and under the microscope they looked like little green footballs. That worked remarkably well.

By the end of June, 2017, the plants had outgrown the length of their skewers. I went to a garden center to buy taller sticks, which then appeared far too long, but who would know? To make sure that our eyes were not accidentally hurt by these sticks while bending down to watch or water the plants, brightly colored Lego bricks were placed on top of each stick as a warning signal. Yet it took only a month for the plants to reach the top of those sticks. By that time, the Lego bricks were removed, because they were no longer needed.

Until the end of July 2017, the pepper plants rapidly kept producing more leaves, branches, and roots, while their main stems were getting thicker and more solid. Clearly, the development of those plants was geared to capturing as much solar energy as possible while the getting was good, while investing that energy into getting more of it.

OUR PEPPER PLANTS JOINTLY MAXIMIZING THE CAPTURE OF SOLAR ENERGY

By the middle of August, 2017, the plants had grown so many leaves that sunlight could no longer pass through them without being captured by them. That observation suddenly evoked a thought. Even though none of the individual plants might have been maximizing the capture of sunlight that fell onto a certain area, their combined emergent effect was exactly that.

This joint maximization of capturing solar energy cannot directly be explained by the theory of natural selection. It is another emergent effect of life. And even though this particular emergent effect may be well known in biology, it was a big revelation for me. Until that time, I had accepted that the theory of natural selection was the major theory for the history of life. But at that moment I realized that it was not. It is a good theory for the origin of species, but it is not sufficiently complete for explaining the entire history of life including all its emergent effects. That bias – in Chapter 1 mentioned as Bias # 2 – was unknown to me until that very moment. Theoretically speaking, this bias had put many people, including me, on the wrong foot.

If Darwin and Wallace's theory of natural selection was not the major theory of the history of life, apparently, we did not yet have such a theory. Could we find one? Further reflecting on these matters, I realized that there were a great many aspects concerning the history of life that did not fit well into the theory of natural selection. How did, for instance, food webs emerge and evolve? And what about the emergent effects of life on geology and the biosphere such as discussed by Vernadsky and Lovelock?

And how would human history fit into such a new scheme, if it could be formulated? Over the years, I had been appalled by the theoretical efforts of scholars who had sought to uncritically apply Darwin and Wallace's theory of natural selection also to human history, as if that theory would also be its major paradigm. Even though aspects such as competition and scarce resources also play a major role in human history, I had felt for a long time that this was not the way to go in terms of a general theory of human history, as if one were trying to push something into a box in which it did not fit very well. Now I began to understand that much like for the history of life, a major theoretical issue of human history were its emergent effects, which could not directly be explained by the theory of natural selection, even though some of them might come as a result of it.

While contemplating all of this, I realized that there were a great many loose theoretical ends in both biological and human history. To some extent, I had already been aware of that, as can be seen in Chapters 5–7 of my book *Big History and the Future of Humanity* (2010, 2015), in which I had sought to deal with them in the best possible way then available. Yet although my account of that period appeared better than other accounts known to me, it still felt unsatisfactory. But at that time I could not do any better, and I had accepted that state of affairs. Now I knew what had gone wrong. I had been seduced into thinking that the theory of natural selection was the general theory of the history of life, while it was not.

I suddenly remembered wonderful conversations with the Belgian biologist Koen Martens, who had lectured in our big history courses for many years in his intelligent, lively, and charming ways. During one of those meetings, about ten years before I began growing pepper plants, Koen Martens told me that Charles Darwin had written a book about earthworms and their effects on the soil. At that time, this did not ring a bell. But I had kept it in mind.

But while further reflecting on all these matters, that reference suddenly jumped up in my mind. I had to find out what was in that book. Thanks to the internet it was very easy to find. On websites such as Darwin Online, one can even access the original edition. I did so, and realized what was mentioned in the previous chapter, namely that Darwin had been aware of some of the emergent effects of life, and that he had not employed his theory of natural selection for explaining them. In that book he did not even wonder how that theory was related to his earthworms' actions. Clearly, the great naturalist had not been hindered by such a bias.

Freeing myself from that bias opened the door to my current thinking. I realized that if there was one such a theoretical bias, there might be more of them. That was the beginning of trying to take distance from the current theoretical thinking about the history of the biosphere, life, and humanity, which opened my mind to my current insights.

This discovery also made me realize that not only were the history of the biosphere and of humanity lacking general theories, but so did the history of life as well. That opened another major floodgate, namely by trying to examine those three theoretical problems jointly, even though those three academic fields appeared almost totally separate. Surely, all those general theories had to be interconnected. A planet gets a biosphere through the joint actions of life. And humanity is a special life-form that emerged within this evolving biosphere. I was not aware of any general theory that connected all of that. Could such a theory be formulated, and if so, how?

MEASURING THE JOINT MAXIMIZATION OF CAPTURING SOLAR ENERGY

On August 17, 2017, while reflecting on the joint maximization of capturing solar energy by our pepper plants, I decided to try and measure that. First, the surfaces of a number of leaves of different sizes were measured, chosen as representatively as possible. Next, the total number of leaves of all our pepper plants was estimated by taking random samples. Those combined data yielded an estimate of the total surface of all pepper plants' leaves combined.

The top surface of the flowerpot was also measured, and subsequently compared with the total surface of all the leaves combined. The result stunned me. It turned out that the total leaf surface was about twenty times as large as the flowerpot's top surface. In other words, the leaves were, in principle, able to capture twenty times as much solar energy as the amount that fell onto the ground in which they were rooted. That unequivocally showed how strong the tendency was of our pepper plants to jointly maximize the capturing of solar energy.[1]

Because the pepper plants were growing vertically, less solar energy was reaching the area around the flowerpot as a result of the shade that the plants caused in those surrounding areas, where, in undomesticated nature, competitors might have been trying to grow and capture solar energy. This immediately explains why many plants tend to grow vertically, why they go into three dimensions, so to speak, because it is a powerful strategy of capturing more solar energy for themselves as soon as the available area they are rooted in becomes a scarce good.

Reflecting on this, it made me realize that humans sometimes do similar things. Constructing tall buildings, for instance, comes as a result of the same basic impulse, namely because the surface area on which these buildings are sitting is becoming scarce. Yet in doing so, our species is not trying to capture more solar energy. Humans are simply trying to make use of the available space in more efficient ways, because our population numbers keep growing while the Earth is not getting any bigger.

To be sure, humans have constructed tall buildings for other reasons as well, first of all perhaps because of prestige, to impress other people. Taller buildings have also been constructed for defense and offence, or to improve the ability to see farther, as well as for being seen better from farther away. Plants do none of that. Because they do not have any brains, they tend to keep things simple. They go into three dimensions mostly, if not exclusively, for capturing solar energy and carbon dioxide.

---

1 For those interested in my measurements, estimates, and calculations on August 17, 2017: the flowerpot had an inner diameter of 28 cm. As a result, its total topsoil surface was 616 cm^2. Based on a partial count of the leaves, I estimated the total number of leaves as about 500 leaves, each of them with an average surface of about 5×5 = about 25 cm^2. As a result, the total surface of all the leaves combined was about 12,500 cm^2. This number divided by the total top soil surface of the flowerpot yielded a factor of 20.29, rounded off here as 20, given the uncertainties in the measurements and estimates.

The pepper plants, August 17, 2017, when their combined leave surface was measured. (Photo by the author)

The situation of our pepper plants would have been different if they had not been supported with sticks. Without them, in undomesticated nature the plants would probably have spread farther out on the ground, and, in doing so, they would have covered more soil surface. That might have lowered the ratio between the total leaf surface and the soil surface that they grew in. Yet more likely than not, in 'nature' they would have faced competitors that were doing similar things. And their combined emergent effects might have led to a joint result similar to the one that I witnessed.

All of that led to the inevitable conclusion that the joint emergent effect of those pepper plants was maximizing the capture of solar energy wherever and whenever possible. And surely, not only our pepper plants were doing that. All of untamed nature was doing the same. This had to be a most important emergent effect of the history of life, and, in consequence, also of the biosphere. Yet I had not read that anywhere formulated in such a way. Had I missed something of importance that others already knew? Or had this simple conclusion perhaps escaped the attention of most, if not all, scholars?

I started consulting recent books on the history of life and its energy uses but did not find such an observation anywhere. I may have missed important literature, and if so, I request the readers to inform me. But with the exception of Vladimir Vernadsky's work, quoted at the beginning of this chapter, none of the rather authoritative overviews of the history of life, and of the biosphere, that I consulted mentioned it. Furthermore, none of the other authors mentioned in Chapter 1 had written about it, while prominent energy scholars of the 1970s and 1980s referenced in the bibliography below, most notably perhaps Earl Cook, Ronald Fox, Engelbert Broda, and Howard T. Odum, had not paid any attention to this phenomenon either.

The only scholar who, to my knowledge, had drawn a similar conclusion was Vladimir Vernadsky, whose work I began to read more closely in 2018 as part of my search for authors who might have said such things. I had not read his work intensively after I had first become aware of him and his work in 2005. At that time, I was working hard on elaborating my general theory of big history explained in my books of 2010 and 2015. Unfortunately, in that respect Vernadsky's book was not very helpful in further clarifying that theoretical scheme. Although also in those years I thought that his book was very interesting, it was a challenging read, not least because the book is not very tightly organized, while it contains a large amount of very densely formulated knowledge. Yet with my new observations and interpretations in mind, it made a great deal of sense to read that book again.

As quoted at the beginning of this chapter, on p. 78 of his book *The Biosphere* Vernadsky mentioned that the 'leaves in meadows at our latitudes are 22 to 38 times larger in area [than the ground they cover].' Furthermore, Vernadsky paid a great deal of attention to how the Earth is covered by green plants wherever possible, which try to capture as much solar energy as possible. That led to a maximum capture of solar energy by green nature. That energy was subsequently used by other organisms, while their actions transformed the planet, thus forming a biosphere.

That was exactly what I realized while growing pepper plants. But Vernadsky had done so about one century earlier. Hat off to this great pioneer! Yet Vernadsky's further elaboration was a little different from mine. His emphasis was not, first of all, on how all of life worked, but rather on its effects within the biosphere. In addition, the illustrious Russian scholar did not systematically consider the history of life including its emergent effects within the biosphere, even though he mentioned a few such long-term developments. All of that is totally understandable, given the lack of sufficient historical data at that time.

Furthermore, Vernadsky tended to see the biosphere in equilibrium. Much depends on what he meant by that, not least because a term such as a 'dynamic steady state,' a relatively stable dynamic situation, did not yet exist in his days. Such a dynamic steady state exists in a great many situations in which the situation as a whole does not change very much, while there is a flux of matter and energy that sustains it. The Earth provides such a case, in which the total amount of solar energy arriving and leaving is about the same. Yet this energy flow produces a dynamic, but relatively stable, situation on its surface.

Being a big historian, I immediately started thinking in terms of a long-term history of the biosphere, with emphasis on living organisms as prominent actors in capturing solar energy and matter, with emergent systemic effects within the biosphere. To be sure, Vernadsky was aware of such processes as well, while the term 'emergent effects' had not yet been coined either. It may therefore be fair to say that while growing pepper plants in 2017, I suddenly found myself on a very similar track as the one that Vernadsky had already been on in 1926. Surely, the great Russian scholar deserves all the credit and honor for his pioneering observations and elaborations.

My effort in this book should, therefore, be seen as an attempt to further elaborate these ideas while integrating them within cosmic evolution / big history. I am not claiming that all my insights presented in the chapters that follow are entirely new. To the contrary, most, if not all, of those ingredients are already well-known. My analysis should therefore perhaps be described as rearranging the well-known furniture in a room while seeking to obtain a more orderly situation. But to explain that in further detail, we first need to pay more attention to what happened next to our pepper plants.

THE SECOND PHASE OF OUR PEPPER PLANTS' SURVIVAL STRATEGY: PREPARING FOR THE LEAN SEASON

By the end of August, 2017, the plants' leaves had further grown and multiplied. According to my rough estimate at that time, the total surface of all the leaves combined had become about thirty times as large as the flowerpot's soil surface. I had not yet studied Vernadsky's work again, including his estimates mentioned at the beginning of this chapter. It is therefore remarkable how close my estimates were to his.

In the meantime, the pepper plants had begun to produce flowers. The first flowers appeared around the middle of July 2017, well before the plants had reached the tops of the supporting sticks. Apparently, after enough solar panels had been put in place and sufficiently strong roots and stems had been made, the plants began to invest in flowers. Such decorative ornaments (as seen from a human perspective) are expensive for a plant. They do not harvest any energy and matter, but instead cost considerable amounts of them. So, why would a plant make flowers? The answer is, of course, to produce offspring in the form of seeds.

Making flowers represented a sudden change in the pepper plants' survival strategy, from a short-term to a longer-term strategy. From the moment the seeds had sprouted until the production of the first flowers, the plants had done everything to achieve immediate results in direct reaction to what was happening in their environment, namely: making leaves to capture sunlight and carbon dioxide as well as roots to capture water and minerals, while the stems connected and supported all of that. Yet making flowers had no immediate use for the plants, while it cost energy and matter to make them.

As we all know, the reason why plants make flowers is to produce seeds that can survive the lean season and start producing new plants during the following favorable season. In other words, making flowers implies an investment in surviving less favorable circumstances that do not yet exist, but which are likely to happen somewhere in the future. In doing so, apparently plants are anticipating a future that is different from their immediate circumstances, while they are acting accordingly by exhibiting a type of behavior that can be called pre-adaptation.

How would our pepper plants have detected that this was the right time to do so? And how would all seed-producing plants have evolved such anticipating survival

Pepper plant flowers, July 29, 2017. (Photo by the author)

strategies? This may be well known, but not to me. In the case of our pepper plants, this change of survival strategy may be related to a first decline of the incoming solar energy. But whatever the causes may have been, the process of natural selection must have favored the emergence and further development of such characteristics among all those species.

The flowers produced pollen, which needed to be transferred to the flowers' stems. The pollen of pepper plant flowers can be spread by insects. That is an efficient way of transporting pollen, and thus saves on the amounts that need to be produced. It is therefore cheaper in terms of energy and matter expenditure to let insects do this job than to rely on the wind to do so, because the delivery of pollen to other flowers is considerably more efficient when done by such little flying helpers than by gusts of air, because pollen carried by the winds are not transported to their targets in efficient ways. In consequence, they may end up anywhere. For airborne pollen to be effective in reaching the reproductive centers of other plants, they must be produced in large quantities, which is costly for the plants.

Yet to attract the insects to the pollen and stems, those flowers need to be attractive to them, both in appearance and by providing an energy and matter reward to those little flying transporters in the form of nectar, while those insects may also eat the pollen themselves. All that added complexity of flowers has an energy and matter price tag attached to it, while for the insects, flowers may be seen as the equivalent of what gas stations (or electrical charging stations today) are for motorists. Those refueling stations need to provide sufficient energy. And they must also clearly be marked as such, so that they can easily be found.

Evolutionary speaking, the spread of pollen by the winds is part of an older survival strategy that would date back to some 380 million years ago. Producing flowers while relying on insects to spread the pollen is much more recent, starting from perhaps around 160 million years ago. Apparently, this more recent survival strategy has paid off, because ever since it emerged, such plants have often out-competed those who stuck to older ways of spreading pollen. Yet even today, many plants that produce airborn pollen have survived to the extent that they create problems in spring for those who are allergic to those pollen.

Another price paid by flower-producing plants is their dependence on insects for their survival. As soon as, for whatever reason, those little flying helpers would no longer come and do their job (as seen from the plants' perspective), those plants are doomed. Here, we see a major theme in biological and human evolution, namely that growing interdependencies tend to create potentially increasing vulnerabilities.

Flowering and pollination are part of a plant survival strategy aimed at diversifying its genes, so as to produce more individual variation. As explained by the theory of natural selection, this allows such organisms greater chances for survival. Because in our Amsterdam apartment there were no insects that could do the pollination, I did it myself using a small paint brush as soon as the first flowers appeared. I had learned this trick in 1980 while working for a year at the ecological farm 'Gaiapolis' near the city of Leyden, where I lived at that time.

Interestingly, pepper plants do not require pollen from other plants. Their flowers can pollinate themselves, by having the pollen moved to their stems. As a result, for their survival pepper plants do not need insects that are bringing in pollen from other plants. But they do need insects, or the winds, for moving the pollen to their flowers' stems. Self-pollination implies that such plants will not achieve as much individual genetic variation as the species that only accept pollen from other plants. But it does improve the chances of producing pepper pods and seeds. Is that the reason why the flowers of pepper plants are small, because attracting insects is not absolutely required for their survival? And by making small flowers, are such plants limiting their energy and matter expenditure? How would all of that have evolved? Would that perhaps be a result of undomesticated pepper plants perhaps growing relatively far away from each other, so that insects might find it hard to go from one such plant to the next? All of this may be well standard knowledge within biology, yet it is unknown to me.

My pollination efforts worked well. Yet not all flowers turned into pepper pods. In fact, about half of them closed their petals after a while and dropped to the ground. Why would our pepper plants do that? Did those unsuccessful flowers need cross-pollination from other plants? Or, while making fertile flowers, is there simply a lot that can go wrong, biochemically speaking, including successful pollination? Was this perhaps part of the plants' 'quality control' that had produced such a high success rate of sprouting seeds? This 50% failure rate clearly increased the cost of making flowers. But by quickly dropping the flowers that did not work, those energy and matter costs were minimized.

How would our pepper plants detect that things were going wrong, and when? After the harvest, we did not find any pepper pods without seeds. Apparently, all the successfully pollinated flowers had produced reasonably effective pods filled with seeds. We did not test, however, whether those new seeds were all fertile. There is no longer a way of knowing that, because after the harvest we discarded the seeds while turning the pods into Peruvian-style pepper sauces. If we wanted to find out, we would need to repeat the experiment.

By the end of August, 2017, the sun was getting noticeably lower in the sky, while the day length was also decreasing. All of that led to a decrease of solar energy available to our pepper plants. Yet their leaves kept growing and multiplying remarkably rapidly, apparently in an effort to keep capturing as much solar energy as possible. Also the stems were growing thicker, as if the plants anticipated that soon they would have to carry the relatively heavy pepper pods. All of that represented a well-coordinated survival strategy, which apparently had been fine-tuned by the process of natural selection.

The plants kept making more flowers, while they began to produce the first pepper pods as well. In contrast to the white and therefore very noticeable flowers, all the emerging little pepper pods were green, a very similar color as the leaves. Is this a form of camouflage so as not to be noticed by animals before the seeds are ripe, while the little green pods may be capturing some solar energy as well?

While further ripening in September, 2017, the growing pepper pods became brightly colored. The Spanish peppers began to turn red, while the little Caribbean peppers turned yellow. Most of the *jalapeños*, however, stayed green. Yet by the end of October, also a few *jalapeños* had turned red. For all of the *jalapeños* to turn red, a longer growing season with more sunlight would have been needed than was available in our apartment. And not all the Spanish peppers turned red either. About one third of them also stayed green.

The pepper plants, September 23, 2017. (Photo by the author)

Why are peppers spicy? The reason appears to be that some animals may not like that taste and would thus not try to eat them, while other animals such as birds would not be affected. That would facilitate the spreading of the seeds by the animals who are not affected by their spiciness, as long as such seeds would survive their trip through the avian intestines. How would such a difference in sensitivity to spiciness have evolved? Would there already have been a difference in such sensitivity between dinosaurs and mammals when both groups of animals were still roaming the Earth together, our current birds being direct descendants of some dinosaurs? In other words, for how long would spiciness have existed and may have functioned as such a selective mechanism?

Whatever the case may be, their spiciness has probably helped the seeds to be carried fast over longer distances. Today, such long-distance dispersal is effectuated by humans, but in our case because some of us like their spiciness, while we do not eat a meal solely consisting of peppers. An efficient dispersal over larger distances is probably advantageous in the tropics and subtropics where wild peppers thrive. In those areas, there is often a great deal of solar energy while the competition for it is fierce, including infective diseases that seek to harvest their energy and matter from the plants. Growing at greater distances from the parent plants may therefore reduce the chances of being attacked by such tiny infectious predators.

Such an explanation ties in well with the earlier reflections on why the survival strategy of pepper plants includes small flowers and self-pollination. Yet in doing so, the seeds must not be digestible by birds, while they would need to survive the trip through their intestines. This puts limits as well as demands on the size and overall structure of the seeds, which explains why pepper seeds are small, hard, and plentiful.

Pepper seeds can survive a lack of water, lower temperatures, and little or no solar energy for at least half a year. I know that for sure, because a portion of our seeds had survived about six months in such a dried state inside our apartment before sowing them. But as soon as the circumstances began to improve, most notably sufficient warmth and humidity, they sprouted. And so the life of pepper plants continues. Every stage of it is geared to survival while adapting to the prevailing circumstances, or to those circumstances that may be coming soon.

HARVESTING THE PEPPER PLANTS

At the beginning of November, 2017, I decided to end the experiment. By that time, the sun, if visible at all, stayed very low in the sky, while the days were becoming very short. Some leaves were getting moldy, while most of the pepper pods had ripened. Even though our daughter had not been very interested in the plants while they were growing, she was happy to harvest the peppers. Was her happiness a type of behavior dating back to the period in which humans lived as gatherers and hunters, during which presumably men hunted animals and women gathered edible plants while taking care of the young children?

Our daughter Giulia harvesting peppers, November 6, 2017. (Photo by the author)

Whatever the case may have been, this was the result of our harvest:

- Thirty-two red Spanish pepper pods, in total: 196 g
- Fourteen green Spanish pepper pods, in total: 71 g
- Twenty mostly green *jalapeño* pods: in total: 160 g
- Four yellow Caribbean peppers pods: in total: 21 g
- Two possibly hybrid pepper pods (both of which contained seeds): 13 g

Those pods yielded at least ten times the number of seeds that had been sown, but probably much more than that, with the exception of the Caribbean peppers, which were the most tropical in nature, and were, perhaps as a result, the least successful within our circumstances.

The weight of all the pepper pods combined was about 460 g. After having taken out the plants, also the total weight was measured of all the leaves, stems, and most of the roots that could be extracted. Their combined weight was 1,235 g. There were some roots that could not be weighed, most of them tiny, because they were mixed in with the soil. In fact, all the soil had become completely caught up by a huge network of fine roots, which made it almost impossible to separate all the roots from the soil. Much like Darwin's earthworms had ploughed the soil, in doing so leading to an emergent effect of life within the biosphere, here we witnessed a similar emergent effect of life that cannot directly be explained by the theory of natural selection.

The first partial pepper harvest, November 5, 2017. (Photo by the author)

My rough estimate of the combined weight of all the leaves, stems, and roots including all those fine roots was a total of about 1,500 g. To my surprise, that was only about three times as much as the combined weight of all the pepper pods. It showed the remarkable efficiency with which these pepper pods had been produced. Apparently, in doing so the pepper plants minimized their investment in their general structure in terms of mass, while maximizing the capture of solar energy, carbon dioxide, and minerals, as well as the production of pepper pods.

As we saw earlier, the plants' leaves were surprisingly light. They appeared to combine a maximum of surface area for capturing sunlight and carbon dioxide with a minimum of investment in its structure, using as little matter as possible, while keeping them in shape with the aid of the accumulated water, under the pressure of which the leaves maintained their form. At the same time, the plants appeared to maximize the production of pepper pods, which was aimed at ensuring a successful next generation. However, we do not know to what extent all those effects may also have come as a result of human selection. Undomesticated pepper pods may be considerably smaller. But possibly more of them may be produced. This would need to be further investigated.

Our pepper pods weighed about ten times as much as the seeds that they contained. Apparently, most of the weight (including a considerable amount of water) is expended on packaging the seeds as well as on making them attractive to birds, which will 'hopefully' take care of their successful dispersal. Grains, by contrast, usually do not do any of that. They tend to defend themselves instead by producing solid husks.

After having measured all of that, we decided to pickle the peppers. The red and green Spanish peppers as well as the little yellow Caribbean peppers jointly ended up in a delicious *ají* sauce, while the *jalapeños* were pickled using a Mexican recipe found on the internet. We did not save any of the seeds. That is how a four billion years' long chain of life was broken, which happens all the time in nature.

But as so often in nature, also in this case only a few members of the species went extinct, but not in any way the entire species. Yet over longer periods of time, things may go differently. According to the US paleontologist David Raup (1933–2015), during the history of life over the past 500 million years, more than 99% of all species that lived during that period are now extinct. This number does not include microorganisms, but only the larger, more complex, life-forms that have left traces in the fossil record. That is a lot of species going down the drain over the course of time.

WHAT ABOUT THE SURVIVAL STRATEGIES OF LIFE?

On the preceding pages, the term 'survival strategy' was frequently used. How original is that term as an analytical tool? On the internet, such a term is frequently used for characterizing specific plants with a meaning that is very similar, if not the same, as employed here. Yet on the pages that follow, the term 'survival strategy' will not only be used for our pepper plants but also for all other living species, from the very simple organisms to the most complex ones, from the tiniest viruses to the largest whales, also including humans.

I may have missed something in the academic literature. But I have not yet encountered the term 'survival strategy' used by biologists or historians in a systematic fashion as part of a general theory of the history of life and humanity. Doing so will be a major focus of this book. In fact, this term has become a cornerstone of my fresh theoretical approach to the biosphere's history. This is also where my theoretical approach begins to deviate from Vladimir Vernadsky's work, because the great Russian scholar did not pay any systematic attention to those aspects of life, even though they fit his approach seamlessly.

In doing so, a further clarification may be needed. My use of the term 'survival strategy' does not imply in any way that all living species have been pursuing their survival strategies consciously. To the contrary, more likely than not, most of these survival strategies are not at all consciously purposeful. Here, the term 'survival strategy' indicates a certain way of doing things that helps to improve that organism's survival chances, no more, no less. In an analogy with today's economic language, we could also call it the 'business model' of a certain species. Yet the term 'survival strategy' may be more acceptable within academia, which is why that has become my preferred term.

When would a differentiation have happened between unconscious survival strategies and consciously purposeful ones, and for which reasons? And how, and to what extent, may unconscious and conscious survival strategies still jointly operate within individuals that are equipped with both? As yet, I do not have good answer to these questions.

Whatever the case may be, it seems clear that for most of life, pursuing their particular survival strategies is virtually all they do. Only a few more complex animals, including humans, may have enough leisure time to do other things than trying to secure their survival, such as reflecting on our biosphere's history. In this respect, people living in more affluent situations may present an exception not only to most of biological history but also to most of human history. For the great majority of people who have lived on this planet, securing their survival has been of overriding importance in their daily lives.

As we saw above, our pepper plants pursued a very specific survival strategy, starting (rather arbitrarily) with seeds during the lean season, followed by plant growth for capturing as much solar energy and carbon dioxide as possible as soon as the circumstances allowed it, while using it for building its structure. This was followed by the production of flowers and seeds to survive the coming lean season. This pepper plant survival strategy is very much tied to the rhythm of their natural environment, while it has continuously been honed by the processes of natural and human selection.

Let us compare this pepper plant survival strategy to what happened to our avocado plants during the following year. In May 2018, in the same flowerpot we planted a few large avocado seeds, all apparently containing considerable amounts of matter and energy. Those seeds came from avocados that had been bought in the same Turkish store around the corner here in Amsterdam. Most seeds sprouted fairly quickly. While making roots, their little stalks shot up quickly to a height of about 10 cm (4 inches) before unfolding their first leaves. By contrast, the pepper plants had already unfolded their first leaves at a height of less than a centimeter. The pepper plants could probably not make such large stalks right away because their seeds were much smaller and, as a result, did not contain enough energy and matter to do so.

Our avocado plants, July 21, 2018. (Photo by the author)

Why did those avocado plants behave so differently? Were they seeking to quickly out-compete lower-growing plants in the struggle to capturing solar energy? The avocado plants could not have sensed the presence of such competitors, at least not in our flowerpot because it did not contain any of them. So, they had all the resources for themselves. Clearly, this avocado survival strategy must have become ingrained through the process of natural selection. More in general, their behavior represents the survival strategy of trees, all of which tend to grow tall to capture solar energy, in the process seeking to out-compete other species for obtaining this resource. This competition must have favored the emergence and evolution of trees.

After getting in place their first solar panels, the avocado plants began to produce more leaves, one by one, while their stems kept growing as well. By the end of July 2018, I decided to take out all the plants except the one that appeared to be the strongest and healthiest. I did so because I feared that to be successful, there might not be enough room in the pot for more than one single plant.

Until the end of November, 2018, that sole survivor of human selection grew taller while producing more leaves. Then it stopped doing so, probably for lack of sunlight and sufficiently high temperatures, because I did not provide any heating to that portion of the apartment during the winter unless there were a serious risk of frost damage. As a result, the average temperature of the loggia over the winter was about 15°C.

By the end of December, the avocado leaves began to wither, starting with the lowest ones. I decided not to do anything and observe instead whether the avocado plant would survive the winter in those circumstances. Sadly, seen from my personal perspective, the avocado plant lost its final leave right before the spring of 2019 finally arrived. I waited until the end of May 2019 to see whether its stem would start making fresh leaves. But it did not. Instead, the little stalk also withered away. That was the end of our experiment. Clearly, this avocado plant had not survived our lean season.

Apparently, the avocado plant's survival strategy was not sufficiently adapted to the annual circumstances in our loggia. The pepper plants, by contrast, would have survived them rather easily, even though the plants themselves would probably have withered away, too. But their little seeds would have been the secret of their success in surviving the winter here, while our avocado plant, which needed several years to become large and mature enough to start making new seeds, did not reach that stage. Clearly, it was adapted to other, more semi-tropical, circumstances.

SURVIVAL STRATEGIES IN THE HISTORY OF LIFE

As noted earlier, all living organisms can be seen as having evolved their particular survival strategies, with the aid of which they position themselves within the struggle for existence. And as explained by the theory of natural selection, their unceasing evolution in doing so has produced the great diversity of life. To some extent, untamed nature can therefore be seen as a neoliberal marketplace for survival strategies, much like how many human markets function today, in which people are seeking to sell certain products. Survival strategies may be rather similar, especially when they work well within certain circumstances. Yet quite often, even those survival strategies are slightly different, in doing so leading to slightly different products, while innovations may sometimes produce rather different products.[2]

All of this great biodiversity occurs as long as the circumstances are sufficiently good. On land, that means first of all enough sunlight, warmth, water, and fertile soil. In the water, as long as it does not freeze or become too hot, sufficient sunlight and minerals are usually the limiting conditions. In the seas and oceans, such good circumstances are usually found relatively close to land, where the water is not too deep, while the water runoff from land brings in the needed minerals. In the middle of oceans, life is often not abundant. The same applies to dry inland areas such as deserts.

The suggestion here is, that for a good understanding of the history of life, and, in consequence, also of the biosphere, we need to systematically consider all the survival strategies that have been pursued by all the life-forms that have existed. This immediately raises the question how many different survival strategies can be distinguished, as well as which criteria would be used to do so. Right now, I do not have detailed answers to those questions, while such an overview still appears to be lacking in the academic literature.

As a result of this situation, we may find ourselves in a situation of considerable confusion and awareness of ignorance. How many survival strategies could be discerned? As many as there are species? Or could they perhaps be grouped into broader categories, much like how species are classified into more general groups according to the classification system first proposed by Carl Linnaeus? If so, would such a classification of survival strategies be similar to the current classification of species? Or would it perhaps sometimes be different, most notably when different species on separate continents evolved similar survival strategies?

None of the biology known to me appears to have posed such questions, while seeking to answer them in a general systematic fashion. Is this only lack of knowledge on my part, or has this theme perhaps escaped the attention of biologists? How could I find out? As yet, I do not have good answers to those questions. But seen from a more general point of view, a classification of major survival strategies appears to be feasible. Those major categories have already been recognized for a long time within academia, yet without calling them survival strategies. This first of all applies to the survival strategies of what in this book will be called the primary captors of energy, both sunlight and geothermal energy, namely microorganisms and plants.

Subsequently, a group of secondary captors of energy can be defined, which comprises those organisms that do not harvest their energy from inorganic nature but instead from other life-forms. Secondary captors include microorganisms and viruses as well as all animals (and also a few plant species). Within this second

[2] I owe the inspiration for writing these sentences to 'our' Dutch paleontologist and paleo-anthropologist John de Vos, affiliated to the Dutch National Museum for Natural History Naturalis in Leyden. For a great many years, John taught his excellent lectures in our big history courses, starting in 1995, when we held our course for the first time. As part of that, we shared a great many stimulating conversations over many beers, from which I learned a great deal, while he also introduced me to colleagues and fascinating places that otherwise would have been inaccessible for me.

category, distinctions could be made between secondary captors that feed on primary captors; that eat the secondary captors; species which are both primary and secondary captors, such as certain plants; and the one single species which does all of that, namely humans.

To be sure, none of that is totally new. But it is formulated here in a way that is slightly different from the ways in which it is usually done. To the best of my knowledge, it is currently not common to speak of primary or secondary captors of energy. The more common terms for such categories are autotrophic and heterotrophic organisms, or primary and secondary producers. Such terms were already employed by Vernadsky.

In this book, by contrast, I want to emphasize the important role that the *capture* of energy by life plays in the biosphere's history. That is why I am opting for those terms. Right now, this may look like nitpicking. Yet in the chapters that follow, I hope to demonstrate why this emphasis on the capture of energy by life is essential for a good understanding of our biosphere's history.

Within those broad general categories of survival strategies, a great many subdivisions can be made, including the differences between the survival strategies of pepper and avocado plants. Exploring that theme in more detail would be a most interesting academic enterprise. Yet that will not be attempted in this book, because doing so would distract us too much from our main subject, an overview of our biosphere's history. As a result, those questions will be left unanswered for the time being.

The main point here is that for a better understanding of the biosphere's history, it is important to think in terms of survival strategies and the associated ways of capturing energy and matter, rather than in terms of individual species, however important that will remain. I am not aware of any academic studies of the biosphere's history that already systematically explore the emergence and evolution of all the general survival strategies over the past four billion years, including their emergent effects within the biosphere.

To avoid any possible misunderstandings: our pepper plants –which came as a result of both natural and human selection– grew in a rather artificial situation. In untamed nature, they would have lived rather different lives, including probably not having a chance to jointly maximize the capture of solar energy, because they might not have lived in the right soil; because they might not have had sufficient access to water and nutrients; because insects and other organisms might have eaten their leaves; and because other plants would have competed with them. The point here is that apparently, all of untamed nature jointly strives toward maximizing the capture of solar and geothermal energy. This will be further elaborated in the next chapter.

Neither do I claim to be an expert on pepper plants. To the contrary, I am well aware of the fact that I am a total novice and amateur in this field. The importance of the observations described in this chapter is not that this is all novel; that this is the way how pepper plants have been behaving in untamed nature; or that I am attempting to instruct the readers about how to grow peppers. None of that! The major point here is, that those observations helped me to get onto the track of formulating more general insights concerning the biosphere's history. This will be further explored in the following chapters.

THE IMPORTANCE OF THINKING IN TERMS OF PROCESSES

Before pursuing all of this in the coming chapters, some attention needs to be paid to the difference between static and dynamic descriptions. Surely, it is important to make a good catalogue of all the species that have existed on this planet in terms of a Linnaean classification. But that is not enough, as Darwin and other realized. For a good understanding of how all of that has worked in the biosphere's history, we need to examine all those species in terms of processes that are continuously interacting. In fact, all of history, from personal histories all the way to cosmic history, can be seen as consisting of interacting processes.

A static description is, for instance, a description of an ecosystem at a certain place and at a certain moment in time. It is like taking a photo of such events. Doing so is rather difficult because there are so many details. How would one decide which details are important and which ones are less so? And how would one explain and understand all of that? By contrast, if one seeks to describe such an ecosystem from the perspective that it forms a process that keeps changing over time, while it is part of, or being influenced by, other processes, then it becomes considerably easier to describe, explain, and understand such situations, even though doing so may appear more daunting at the beginning of such an investigation. This is the case form both biological and human history.

For instance, in the 1980s and 1990s I studied Andean religious festivals not as fixed entities but instead as processes, while they were embedded in larger social and ecological circumstances, which could also be seen as processes. And if one included in the analysis the history of such festivals over many years (if that is what happened), such events may become even more interesting and better understandable. To be sure, doing so requires much more work, not least because one needs to obtain sufficient data to produce such an analysis. But the result may lead to a far greater understanding of such festivals, and may, in consequence, be far more rewarding and satisfying. In Chapter 3 of my book *San Nicolás de Zurite* (1995), a process analysis is offered of the history of the problematic patron saint's festival over centuries as it was realized in that Peruvian Andean village.

Doing so was part of how I learned to understand the history and present of Peruvian religious politics in terms of processes. This general approach, in terms of the sociology of Norbert Elias, became very valuable and useful knowledge. Later, I applied the process approach also to describing and analyzing big history, while in this book, the biosphere's history will be similarly interpreted. The art of writing history in terms of processes will receive more attention in Chapter 4.

In this respect, Vernadsky's work provides a combination of rather static descriptions of the biosphere together with a few process analyses. For lack of historical data, the eminent Russian scholar could hardly have done better. But as a result of this mixture of static and dynamic descriptions, his work lacks theoretical coherence.

So where do we go from here toward formulating a fresh theory of the biosphere's history?

BIBLIOGRAPHY

Broda, Engelbert. 1978. *The Evolution of Bioenergetic Processes* (revised reprint). Oxford & New York, Pergamon Press.

Cook, Earl. 1976. *Man, Energy, Society*. San Francisco, CA, W. H. Freeman & Co.

Fox, Ronald F. 1988. *Energy and the Evolution of Life*. New York, W. H. Freeman & Co.

Howard, Thomas Odum. 1971. *Environment, Power and Society*. New York, John Wiley & Sons, Inc.

Raup, David M. 1993. *Extinction: Bad Genes or Bad Luck?* Oxford & New York, Oxford University Press.

Spier, Fred. 1994. *Religious Regimes in Peru: Religion and State Development in a Long-Term Perspective and the Effects in the Andean Village of Zurite*. Amsterdam, Amsterdam University Press. http://www.bighistory.info/bhi_005_035.htm.

Spier, Fred. 1995. *San Nicolás de Zurite: Religion and Daily Life of an Andean Village in a Changing World*. Amsterdam, VU University Press. http://www.bighistory.info/bhi_005_035.htm.

Spier, Fred. 2010. *Big History and the Future of Humanity*. Oxford, Wiley-Blackwell.

Spier, Fred. 2015. *Big History and the Future of Humanity, Second Edition*. Oxford, Wiley-Blackwell.

Vernadsky, Vladimir I. 1998. *The Biosphere*. New York, Copernicus Springer Verlag (1926, Original Version in Russian).

3 Examining Key Concepts of Our Biosphere's History

> Energy is the only universal currency: one of its many forms must be transformed to another in order for stars to shine, planets to rotate, plants to grow, and civilizations to evolve. Recognitions of this universality was one of the great achievements of nineteenth-century science, but, surprisingly, this recognition has not led to comprehensive, systematic studies that view our world through the prism of energy.
>
> Vaclav Smil, *Energies: An Illustrated Guide to the Biosphere and Civilization* (1999, p. x)

CAPTURING ENERGY AND MATTER SHAPES LIVING BODIES TO A LARGE EXTENT

Before discussing a novel version of the biosphere's history, we first need to consider a few key concepts in more detail. In doing so, let us begin by examining the rather common perception that the major thing living organisms do, is to build and maintain their bodies, and that this is the reason why organisms need to eat, which is considered a secondary aspect of life. Such a view is, for instance, expressed in biology by calling organisms that capture inorganic energy the primary producers (of bodily complexity), while life-forms that eat others are called secondary producers.

Surely, all life-forms need to shape and maintain their bodies. But we also need to examine the undeniable reality that while doing so, all those living bodies are very much geared to obtaining sufficient energy and matter. Doing that is a most important aspect of the daily survival strategies of all living organisms. As a result, their bodies have been shaped, and honed, by the process of natural selection so that they can capture sufficient energy and matter, at least in principle. Without doing that, and, as a result, not having such body shapes, living beings would not exist. The bias of maintaining bodily complexity as the major goal, without sufficiently considering the important role that capturing energy and matter plays in shaping those bodies, will be called Bias # 3 in this book. It is a very common bias in accounts of the history of life.

While watching our pepper plants grow, I became aware of this bias. Capturing energy and matter is the major reason why those plants made so many leaves, which explains to a large extent why they look the way they do. This is the case for all organisms. It explains why sharks have large jaws as well as strong and fast bodies that are able to capture their prey quickly and eat it efficiently. It also explains why our heads are loaded with sensors around our mouth, most notably our eyes, ears,

DOI: 10.1201/9781003275350-3

and nose, so that we can efficiently find the food, assess it, and get it into our bodies. And so the list goes on.

Of course, bodies are also shaped by other necessities of life, depending on the organism, such as digesting the food and using it for building and maintaining their bodies, for moving around, for reproduction, etc. All those aspects of bodies are also honed by the process of natural selection, and they have been receiving a great deal of attention in biology. Yet it remains undeniable, or so it seems to me, that capturing energy and matter shapes living bodies to a remarkable extent, and that this aspect, as a most important part of life's survival strategies, may not yet have received the theoretical attention that is deserves.

Where would such a bias come from? My preliminary answers to that question are as follows. In the first place, we observe all organisms, alive or dead, through their bodies or their fossil remains. This makes it very seductive to focus the attention on those aspects. Yet it is the urge for survival, the urge to maintain that bodily complexity and to pass it on to the next generation, as was argued in the preceding chapter, that provides the main driving force for keeping all of life going. And that survival urge implies the imperative need for capturing enough energy and matter to keep going, which has shaped all those bodies to a large extent.

In the second place, many scholars, most notably perhaps tenured academics living in affluent societies, usually expend little time and energy on obtaining their food, while their incomes appear guaranteed. As a result, such scholars may underestimate the importance of capturing energy and matter to keep going for the history of life and humanity.

And in the third place, for many people living in today's affluent societies energy is cheap, while they may perceive it as something that facilitates us to do things: filling up your car with gas or electricity to go somewhere; turning on a light, or a stove, or one's computer and cell phone. Doing all of that has clear purposes, and it can be done by just plugging in a cord or flipping a switch. Such things are possible today, because many of us live in situations in which concentrated forms of energy have become cheaper and more readily available than ever before, which is all part of the industrial revolution as it took off in the second half of the eighteenth century.

However correct such a view may be to understand the current situation, it would be incorrect to underestimate the role energy plays in all of that, and even less so, to project such views back onto the entire history of life and humanity, because in earlier days, that was not the case at all. To capture the usually less concentrated energy flows occurring in nature such as sunlight and the resulting wind and water flows, humans had to work hard in agriculture, or had to build and maintain wind and water mills or sailing ships. Those people would have known that capturing energy is essential for doing what they needed or wanted to do, and that they had to perform a considerable amount of work to get that done.

The need for energy to maintain greater complexity does not solely occur in the history of life. More generally, it happens in all of cosmic evolution / big history. Many forms of complexity in the universe may gulp up energy and matter from time to time. Black holes, for instance, may devour entire stars, while our Sun regularly swallows comets. All of that results from the gravity that these large bodies exert. They do not try to capture such celestial bodies actively using an elaborate mechanism. And without consuming stars or comets for a while, they would not experience an urge to start doing so, a feeling of hunger for stars or comets. Yet all of life does experience hunger in one way or the other, which is a biochemically-based stimulus that tells organisms that they are running low on energy and matter, and, in doing so, tells them to go and get it. That hunger stimulus, present in all life-forms, may not yet have received the theoretical attention it deserves either.

WHAT ABOUT FREE ENERGY?

Because it is imperative for life to capture energy to keep itself going, we need to pay some more attention to the type of energy that is being harvested. In science, this form of energy is known as 'free energy.' Why is it called that? To understand that, we need to delve a little into the science of thermodynamics. As mentioned in Chapter 1, this branch of science emerged in the nineteenth century after people had begun to use steam engines on a large scale.

Those steam engines were basically water boilers heated by a coal fire. The resulting steam pressure was fed through a pipe to a cylinder where it pushed a piston. This movement was turned into a rotary motion, a spinning wheel, that could perform the desired work: pump up water in a mine; power machines in factories that performed spinning and weaving; make steam trains and ships go; and also, a little later, generate electricity.

Model steam engine 'Made in Western Germany' that we played with as a kid. (In our possession, photo by the author)

A major issue at the time was how efficient such steam engines could be made, how much energy present in coal could be converted into useful work. That was an urgent question, because engineers had noticed that not all the energy contained in coal could be turned into useful power. So, how did that work? What would theoretically be possible, and what could be achieved in practice?

The scientific enquiries by great pioneers such as the French scientist and engineer Nicolas Carnot (1796–1832), the German physicist Rudolf Clausius (1822–1888), the Austrian physicist Ludwig Bolzmann (1844–1906), and the US scientist Josiah Gibbs (1839–1903), led, among other things, to the concept of 'free energy.' That is the energy able to perform work, to make the wheel of a steam engine spin. Furthermore, it was discovered that the ability of steam engines to do so efficiently very much depended on the ambient temperature. Paradoxically perhaps, the colder the outside temperature was, the more efficient the steam engines could run as long as the water boiler was sufficiently insulated and other heat losses were also minimized.

Apparently, the difference between the ambient temperature and that of the boiling water was essential in determining a steam engine's efficiency. This meant that during the winter, steam engines could run more efficiently, at least in principle, than in the middle of the summer. And none of them would work if the outside temperature were the same, or even higher, than that of the boiling water. Such situations would, of course, never occur in practice.

A few important more general conclusions also emerged from thermodynamics. As mentioned in Chapter 1, this theory turned out to be not solely applicable to steam engines, but, in fact, also to the entire universe and everything in it. The reason for that is that every form of greater complexity needs an energy flow. Stars such as our Sun, for instance, exist by converting nuclear energy into heat and radiation. That is what makes stars shine. And life on Earth, including our pepper plants, harvests some of that solar free energy to keep going.

In all these processes, higher-value energy is converted into lower-value energy. In the case of steam engines, a part of it is used to perform work. Sunlight cannot perform any work on the Sun's surface, but the solar radiation that reaches Earth can do so here. The major reason why that is possible is that the temperature difference between the Sun's surface – about 5,500°C, which is, in consequence, the 'temperature' of solar radiation – is much higher than that of our planet's surface, which is, on average, only about 15°C. So, one could say that sunlight carries the temperature of the Sun's surface to Earth. That is a large temperature difference, much larger than the one between steam engines and their environment. As a result, solar energy can, in principle, perform a considerable amount of work within the biosphere.

Our planet also radiates light out into space. But because Earth's surface is so much cooler than that of the Sun, that outgoing radiation is also much cooler. Our eyes cannot see that outgoing radiation. It is therefore called infrared, 'below red,' light, the color red being the longest wavelength of light that our eyes can see. The reason why we can see the incoming sunlight but not the outgoing 'earthlight' is that our eyes have evolved, under the pressure of natural selection, to see everything that is illuminated by sunlight, because that conveys useful information to us, most notably where and how to get food as well as how to avoid danger. Yet it is possible to make that 'earthlight' visible by using special instruments such as night-vision goggles.

Compared to sunlight, earthlight contains much less energy per light particle. Those particles are called photons. And because our planet's surface is not heating up very much – with the exception of climate change – the total amount of energy reaching Earth by sunlight must roughly be equal to that of the outgoing earthlight. This means that earthlight must contain many more photons (but each of them with a much lower energy content) than the incoming sunlight.

Much like how sunlight carries the solar surface temperature in its photons, so does earthlight with our planet's surface temperature. As a result, no work can be performed by earthlight on the surface of our planet, much like sunlight cannot perform any work on the Sun's surface. Because of this situation, earthlight cannot be captured and used by plants for making energy-rich molecules. Instead, all living organisms, plants and humans included, radiate out a certain amount of this low-level energy. That is how we keep our energy budgets balanced, which explains why we do not heat up too much as a result of all the food that we eat.

While considering all of that, 19th century scientists concluded that there are forms of energy that are able to do work, called 'free energy,' while other forms of energy cannot do so, or to a lesser extent, and that this very much depended on the prevailing circumstances. Yet regardless of all these energy transformations, it also turned out that the total amount of energy within a closed and stable system remains constant. To be sure, within such a system, the amount of free energy can, and usually will, decrease. Yet interestingly, the total amount of energy will remain the same (the conversion of matter into energy through nuclear processes excepted). This discovery became known as the First Law of Thermodynamics. This law states that the total amount of energy within a closed and stable system remains constant.

That law would, in principle, also apply to the universe as a whole (nuclear conversion processes excepted). That is what 19th century scientists thought after thermodynamics had been formulated. However, in the twentieth century it was discovered that the universe is expanding. This means that from a thermodynamic point of view, the universe is not a stable system. As a result, classical thermodynamics cannot directly be applied to the history of the universe. In seeking to address this issue, the US astrophysicist Eric Chaisson has made valiant contributions. Yet it seems to me that more rethinking is needed for adapting classical thermodynamics to the dynamic cosmic history as we understand it today.

Fortunately, for studying the biosphere's history, which comprises only a very small fraction of the universe, this hardly makes a difference. When the Earth emerged about 4.6 billion years ago, the universe was already more than nine billion years old, while during that previous history, it had already become very large and mostly cold. Over the past 4.6 billion years of our planet's existence, the universe has kept expanding, while our Solar System and galaxy have not done so.

As a result, a reasonably good first thermodynamic approximation of the biosphere's history is, that this history has taken place within a reasonably stable cosmic situation, which is open, but not noticeably expanding, seen from a biospheric point of view. That does not mean that there were no cosmic changes during that period which would have influenced Earth. But that is a different problem, which will receive more attention in the coming chapters. What matters here is that we can apply classical thermodynamics reasonably well to the biosphere's history.

Yet even though the total amount of energy remains constant within a closed and stable system, the qualities of the various types of energy that it may contain do not. Because of the tendency for higher-value energy (free energy) to be converted into lower-value energy, over the course of time our cosmic environment, and also the universe as a whole, will run out of free energy, much like a car will run out of fuel after a while. Why would that be, and how does that work? And what would be some major consequences?

THE SECOND AND THIRD LAW OF THERMODYNAMICS

The fundamental insight that free energy in the universe is going down, and that, in the process, all forms of energy are becoming more similar, led to the idea of a cosmic heat death, namely that at a certain point in time, there would not be any free energy left able to do work. At such a moment, everything would have acquired the same low temperature. That is why it was called the heat death of the universe. When this idea was formulated during the second half of the nineteenth century, no one knew yet that the universe was expanding, which throws a bit of a spanner into such arguments. Yet this cosmic expansion does not contradict the idea that free energy is running down over the course of time, and that, as a result, the universe tends toward a heat death.

In other words, over the course of time the universe is moving toward a situation of fewer differences and more similarities. Apparently, if left to itself, the universe will become more homogeneous. This line of thought led to the concept of entropy. This difficult word is simply a measure for the lack of structure within a certain situation. The more homogeneous a situation is, the more entropy it contains. By contrast, a highly-organized living body contains far less entropy, because it consists of very complex forms of molecular organization.

How can we explain the concept of entropy in simple terms? In my lectures, I usually do this as follows. Imagine that we would put all the different chemical substances that our body consists of in a number of separate vials, many tens of thousands of such little bottles, that jointly contain our entire body, yet with a lower level of organization. Would we ever be able to reconstitute a human body by putting all these different molecules together again? Obviously, the chances of that happening successfully would be very close to zero. This immediately indicates how much higher the level of molecular organization is within our bodies compared to the sum of all the molecules separated in those vials. Formulated in thermodynamic language, this means that our body contains far more organization, and, in consequence, far less entropy, than all the chemicals in those little bottles.

Yet apparently, the universe is running down on free energy. This means that overall, it is becoming less organized and complex, while the total sum of its entropy is increasing relentlessly, in doing so tending toward a maximum. Only in small pockets such as stars and planets do forms of greater complexity exist, which, in consequence, contain less entropy. But the price paid for this local rise in complexity is that elsewhere in the universe, entropy increases even more.

The tendency for entropy to increase toward a maximum, which is reached when everything has become homogeneous, is known as the Second Law of Thermodynamics. The Austrian physicist Ludwig Boltzmann explained this mathematically by showing that all matter and energy tend to disperse toward their statistically most likely situation. As my university teacher Dr. Joustra explained in 1972 while taking a thermodynamics course at Leyden University, it is extremely unlikely that all air molecules in a room will suddenly converge into one small area. Instead, they tend to be distributed very evenly across the room, simply because that is their most likely distribution. Boltzmann's pioneering approach is known as statistical thermodynamics.

So, in sum: while the First Law of Thermodynamics states that the total amount of energy is constant, the Second Law states that entropy tends toward a maximum, because that is the most likely distribution of energy and matter.

What would such an end situation look like? Seen from a classical thermodynamic point of view (without an expanding universe), after a very long period of time, the universe would reach a stable, very low, temperature, close to zero degrees Kelvin. This is the coldest possible temperature, because at that temperature all the movements of the all the atoms come to a halt. Within this context, it is important to know that what we call temperature is, in fact, a joint emergent effect of the all movements of all atoms and molecules combined. The absence of such movements implies, therefore, the lowest-possible temperature.

As soon as that lowest point has been reached, entropy will have reached its maximum. The tendency of the universe as a whole to move to such a final state is the subject of the Third Law of Thermodynamics. This law is not directly relevant for a consideration of the biosphere's history, because in all likelihood our biosphere will have ceased to exist long before the universe reaches its end stage. But within the context of this book, it is important to know what those three laws are, as well as that there are currently no more than those three laws of thermodynamics, because in the next section, a Fourth Law of Thermodynamics will be proposed, and later in the book, also a Fifth Law of Thermodynamics.

A FOURTH LAW OF THERMODYNAMICS?

How would life's actions fit this theoretical scheme? As we saw earlier, life exhibits a tendency toward jointly maximizing the capture of the available solar free energy. But life is also capturing geothermal free energy, which leads to the same tendency. As a result, it can be stated that all life-forms combined are jointly striving toward maximizing the capture of all the available free energy.

How can we view this from an historical perspective? We do not know whether life originated on our planet or not. Yet scientists agree on the idea that the first organisms on Earth must have been very simple, and that they kept themselves going by harvesting geothermal energy. They did not yet capture solar energy, because doing so requires a more complex biochemistry. As a result, the first period during which life would have tended to jointly maximize the capture of free energy must have involved harvesting geothermal energy. Today, we find microorganisms doing so whenever and wherever possible, while many of them appear to be descendants of those ancient bacteria. Such microorganisms are known as *Archaea*, the 'ancients' in Greek. This situation today clearly exhibits the tendency of jointly seeking to maximize the capture of geothermal free energy wherever possible.

Over the course of time, some of those microorganisms invented ways to capture solar energy. This would have happened first in the water, and later on land. That led to a similar tendency toward jointly maximizing the capture of free energy, but in this case electromagnetic, solar, energy. In our account, all life-forms that harvest geothermal and solar free energy will be called primary captors of free energy. This will be further discussed in the coming chapters.

The important point here is that there has been a tendency for life to jointly strive toward maximizing the capture of all the available free energy, because when successful in doing so, such organisms tend to expand in population size while spreading around. This has depended both on the prevailing circumstances and on the biochemistry that those life-forms had evolved. This tendency for jointly striving toward maximizing the capture of free energy is, in my opinion, the most important emergent effect of life, because it has driven all the further changes that turned our planet's upper layers into a biosphere.

Has this tendency ever reversed or broken down? While individuals may decide to end their lives, sometimes even in small groups, no species as a whole ever appears to have decided that it was enough, and that from now on, they would stop harvesting free energy and matter. To the contrary, during the history of life, the opposite appears to have happened, namely life tenaciously clinging on to its existence, even when the circumstances were becoming worse. And even after some species go extinct, others tend to take their place in jointly striving toward maximizing the capture of the available free energy.

Furthermore, all life-forms exhibit the tendency to multiply beyond their reproduction numbers, while the planet does not get any bigger. Over the course of time, this has inevitably led to life's expansion around the planet, and, in doing so, to jointly strive toward maximizing the harvesting of the available free energy.

In consequence, life's tendency to do so appears to be as irreversible as the tendency of entropy to increase, as expressed in the Second Law of Thermodynamics. That is why I am now going to be bold, perhaps irresponsibly so, and state that the tendency of life to jointly strive toward maximizing the capture of the available free energy constitutes another basic law of thermodynamics, which, from now on, will be called the Fourth Law of Thermodynamics.

The Fourth Law in action in our Amsterdam neighborhood, End of Sporenburg, June 18, 2019. (Photo by the author)

In fact, Vladimir Vernadsky had already noticed such a tendency, which he called the pressure of life, first mentioned as such on p. 59 of his book *The Biosphere*. Yet to my knowledge, the Russian scientist never took the step to recognize this tendency as a major law of thermodynamics.

WHAT ABOUT SECONDARY AND TERTIARY CAPTORS OF FREE ENERGY?

Until now we have mostly been looking at the primary captors of geothermal and solar free energy. Yet as we all know, over the course of time a great many species have emerged that do not harvest such energy, but rather prefer to capture the free energy accumulated within other life-forms. That is a very different survival strategy. Such captors of free energy range from very tiny life-forms, such as microorganisms and viruses, to the largest animals the world has ever seen. All of them have evolved their own specific survival strategies, yet they all share the same general characteristic, namely capturing the free energy and matter that is available within other living organisms. When feeding on primary captors, they will be called secondary captors. And when they eat secondary captors, they will be called tertiary captors.

Because also secondary and tertiary captors tend to expand in population size whenever and wherever they are successful, also their joint emergent effects tend toward maximizing the capture of free energy. In other words, the Fourth Law of Thermodynamics also applies to them. The joint effect of secondary and tertiary captors is not, however, to step up the maximum amount of free energy that is captured by all of life, but instead jointly using it more efficiently. Instead of plants decaying, for instance, and, in doing so, releasing whatever free energy is left in them, they are eaten by other organisms that thrive on them. And such organisms may again be eaten by others, etc. All of that leads to a more efficient use by all species combined of the available free energy originally harvested by the first captors.

One might object that by tapping free energy from the primary captors, the secondary captors put a brake on the primary captors' efforts to harvest geothermal and solar free energy. And that is correct. But for two reasons, that does not invalidate the Fourth Law. The first reason is that the Fourth Law does not state that the capture of free energy is maximized, but instead that life jointly strives toward maximizing it. As soon as the circumstances get worse, the amount of free energy jointly captured will go down as well.

The second reason is that as soon as the capacity for capturing free energy of one single species goes down because it gets eaten, other species may soon take its place, in doing so again stepping up the joint capture of inorganic free energy. The result of this situation is an increasing biodiversity of species. A direct consequence of this situation is that over the course of time, a 'trophic web' emerged, consisting of an increasing variety of organisms. All of that is all is, in principle, well-known. Yet I do not know any biological analysis that seeks to answer the question of how all of that theoretically fits both the theory of natural selection and the biosphere's history.

To be sure, pursuing such an approach is complicated and difficult, especially if we want to analyze today's situation. But that is where the process approach comes to our rescue. By examining step-by-step how all of that has emerged and evolved over time, it becomes much easier to understand all these changes as part of historical processes. My first attempt at formulating such a process analysis will be offered in the Chapters 5 and 7. In this chapter, we will instead focus on general considerations, including two more biases that may stand in the way of such a process analysis.

But before doing that, it is important to see that also the term 'trophic web' may contain Bias # 3 mentioned earlier, the tendency to focus on the effects –in this case: eating food– rather than on the efforts of capturing it. As a result, instead of 'trophic web' and trophic pyramid,' in this book the terms 'food-capturing web' and 'food-capturing pyramid' will be used, in which 'food' stands for the free energy and matter captured by life.

BIAS # 4: FOCUSING ON MORE COMPLEX SPECIES

Before continuing our examination of fundamental concepts, let us first consider another bias that is rather common in accounts of biological evolution, namely focusing on more complex, more spectacular, secondary captors, while neglecting most other life-forms, including most, if not all, the primary captors of free energy, especially those that are little.

Although attention for primary captors and tiny species has not been lacking, I have not yet encountered even one systematic account of the history of life that emphasizes and elaborates the idea that this history is none other than that of the entire food-capturing web including all its emergent effects. The tendency to pay attention to more complex species in biological evolution instead of examining the entire food-capturing web will be called Bias # 4 in this book.[1]

For those who think that this view is not correct, let me give an example, by considering the so-called Cambrian explosion of life-forms. This period started around 540 million years ago, when for the first time in our biosphere's history, a large proliferation of complex life rather suddenly emerged. Why is a great deal of attention on the Cambrian period usually focused on the emergence of animals, one may wonder, while usually very little is said, by comparison, about the emergence and evolution of plants, the first complex primary captors of free energy, from which all those similarly novel and spectacular animals must have captured their free energy?

In other words, what about the emergence of that first complex food-capturing pyramid as a whole and its emergent effects? If humans had been plants instead of animals, would we perhaps have focused our attention more on those primary captors of free energy? In other words, are we paying more attention to animals because we are animals ourselves? Are we, in doing so, implicitly, and perhaps instinctively, seeking to trace our own history back to those early animals? And what about the dinosaurs? Why is there so much more attention for those animals rather than for the species that they were eating, especially the primary captors of those times that

[1] In his book *Holistic Darwinism* (2005) Peter Corning shows a keen eye for the complexity of development of life as a whole. Yet he hardly pays any systematic attention to the interactions between life and geology, while he mentions the biosphere only in passing. Furthermore, he does not appear to be familiar with Vernadsky's work. Many thanks to Clément Vidal for making me aware of this study.

kept their contemporary food-capturing pyramid going, including those spectacular animals?

While analyzing the history of life, this bias becomes even more pronounced as soon as biological evolution led to the emergence of early humans. In many accounts, especially those of cosmic evolution / big history, it is almost as if from that moment on, the rest of living nature becomes the 'natural environment' instead of an integral part of the history of life and humanity. In my book *Big History and the Future of Humanity*, I tried to counter that trend as much as I could. But also in that book this bias is still visible, even though while writing it, I felt something was not right, while seeking to do better. Yet at the time of writing that book, I did not yet know well enough how to handle that issue.

Yet if one looks at all of that from a distance, let us say from lunar orbit, there can be no doubt that if we want to provide a good narrative of the history of life, we must consider the history of the entire food-capturing web: of all the organisms involved, all their relationships, their survival strategies, as well as all the emergent effects with the biosphere. That is the challenge academics are facing if we want to understand our biosphere's history.

While considering this academic state of affairs, one realizes that again we find ourselves in a situation of considerable confusion and awareness of ignorance. We may possess a great many pieces of this gigantic puzzle. Yet to my knowledge, they still need to be put together. Within this field, Vladimir Vernadsky is still the great pioneer, even though his book was published almost one hundred years ago, because the great Russian scholar already saw so much of that. Yet he did not try to systematically sketch a history of the biosphere. His attention was mostly focused on the influence of life on geology, and much less on the dynamics of the history of the food-capturing web.

Someone may object that all of this is totally obvious. So, what would be the big deal here? My answer would be that if all of this is indeed so obvious, why has it not yet become standard knowledge? Why would we not find such integrated accounts in our textbooks? Apparently, that is not the case, because it is not yet obvious to most scholars. A similar objection could have been made against Darwin and Wallace's theory of natural selection. It seems pretty obvious in hindsight, doesn't it? But who thought about it before they did?

Here we encounter a situation that is fairly common in academia. It may be very difficult to find fresh points of view while most academics are looking into other, more established, directions, even though such ideas may be staring in one's face as soon as one begins to look at them in different ways. As soon as one's mindset has shifted a little and starts paying attention to things that appear to fall outside of the established frames of thinking, such fresh yet simple views may suddenly appear. There are many examples of such a situation, including Isaac Newton's theory of gravity and James Watson and Francis Crick's discovery of the structure of DNA. After having been discovered, others may see those ideas as well. Some of them may think all of that is obvious, while they may wonder why they themselves, or others, had not thought of that before.

As soon as such fresh perceptions represent reality sufficiently well, they may become standard knowledge, to the extent that almost everybody may forget who first came up with those ideas. That happened, for instance, to many of the ideas launched by the great Prussian naturalist Alexander von Humboldt, who is my academic hero (he was Charles Darwin's academic hero, too). Today, while looking at maps showing distributions of plants, animals, and climates, who would know that we are, in fact, looking at his pioneering work? And this is only one of his great many contributions to science.

And who would know today, that the idea of a weather forecast as well as the word 'weather forecast' itself were coined by no other than the British naval officer Robert Fitz-Roy (1805–1865), captain of the Beagle that took Charles Darwin on his five-year long voyage around the world?[2] While Darwin's account of this voyage has become famous, Fitz-Roy's account is, in my opinion, similarly fascinating. Like Darwin, also captain Fitz-Roy was a most intelligent man with wide-ranging interests, who could observe and describe aspects of reality extraordinarily well.

Perhaps it is a compliment when fresh ideas turn into virtually anonymous standard knowledge. Perhaps this may also result from the fact that there were no spectacular new general theories behind such ideas. But such a situation implies a bias toward academics who have come up with fresh theories, while other ideas that may have been similarly influential receive less personal recognition, if at all.

WHY WOULD THE FOURTH LAW OF THERMODYNAMICS BE IMPORTANT?

Why would this simple idea, the tendency of life to jointly strive toward maximizing the capture of free energy, be considered so important as to call it a law of thermodynamics? The answer is simple. This drives the history of life, because that is what drives all these further survival strategies. What could be more important than that, while considering the history of all species that have lived on our planet and their emergent effects on it?

But there is more to it. This tendency is also the primordial cause of all the major interdependencies among all the species as part of the food-capturing web that they form with each other. There are also other causes for interdependencies among species, such as, for instance, why many birds like to build their nests in trees. More about that in the following chapters. But the principal dependencies among living species are related to the major issues of life, namely the problems of finding food and avoid becoming food, as William McNeill formulated on page five of his book *Plagues and Peoples* (1976). What the great world historian saw as a central problem in human history is very similarly also the case for the history of life as a whole. In addition, it is the capture of free energy that makes possible all the emergent effects of life within the biosphere. It allows earthworms to plow the soil, as Darwin formulated it, or any other actions undertaken by life, including the transformation of the biosphere currently effected by our own species.

---

[2] I do not know why, Charles Darwin in his *Autobiography* (1958) and Alexander von Humboldt excepted (2004, p. 483), almost everybody else has spelled the captain's last name as FitzRoy of Fitzroy, while in his own account, the captain himself consequently spelled it as Fitz-Roy, including on the title page of his book of 1839.

Why would life seek to capture so much free energy? There are several answers to that question, perhaps even more than what I can see right now. First of all, life needs to survive periods of insufficient availability of free energy, such as winters, as well as periods during which life cannot do anything with the available free energy, such as dry seasons. In other words, the capture of free energy by life is facilitated, and limited, by the prevailing circumstances as they exist within the biosphere.

In the second place, because life's complexity is inherently fragile, for its continued survival it needs to produce a next generation. And the more offspring it produces, the better its survival chances are, as a species. Doing all of that requires large amounts of free energy. And in the third place, because we live on a planet with limited resources, the more individuals there are, the more intense the competition for those resources will become. Here we find ourselves again in the domain of the theory of natural selection, which drives living organisms to seek harvesting more energy and matter, as well as to do so more efficiently.

WHAT ABOUT FLOWERS?

Formulating the Fourth Law as the tendency to jointly strive toward maximizing the capture of free energy made me wonder whether a similar tendency existed for flowers that require pollination by insects, namely: jointly striving toward maximizing the covered area in their attempts to maximize the chances of attracting their little winged friends. To some extent, this may be the case.

Yet such maximizing efforts are limited by the need to power all of that by capturing solar energy. In other words: as soon as flowers jointly covered all the available space, the leaves could no longer harvest solar energy, and the plants would die as a result. This poses a clear limit on the amounts of flowers plants can make. It is therefore not surprising that as a consequence of this situation, plants have responded by producing an increasing variety of flower shapes, sizes, and colors.

This situation also helps to explain why our pepper plants only have a relatively small number of little, mostly white, flowers. Because they are self-pollinating, and thus stand a greater chance of fertilizing their flowers and form seeds through the actions of insects or the winds, they need to attract fewer flying friends to survive than plants that cross-pollinate. In consequence, to attract sufficient numbers of insects to guarantee their survival, pepper plants do not need to invest in great numbers of flowers that are larger, more complex, and colorful.

WHY HAS THE BIOSPHERE'S COMPLEXITY BEEN INCREASING AT AN ACCELERATING RATE?

The Fourth Law immediately explains why the biosphere's complexity has been increasing over the course of time as well as why this increase has been accelerating. Because life-forms compete in seeking to capture enough free energy and matter to survive and prosper, biological evolution has led to increasing numbers of species doing so, in increasingly varied ways, while increasingly modifying the biosphere through those actions.

All of that has led to complex interacting feedback processes within the biosphere, which jointly have been producing unceasing change, favoring some organisms while damaging others. Over the course of time, the combined effect has been an increasing biodiversity which jointly seeks to capture ever greater portions of the available free energy and matter. And because an increasing biodiversity implies more competition for scarce resources, this led to novel ways of doing so through the process of natural selection. Over the course of time, with ups and downs, this has produced an acceleration of all these changes.

In other words, under pressure of the Fourth Law of Thermodynamics, the biosphere has acquired an inbuilt tendency to become more diverse while these changes have been accelerating. This will be elaborated in Chapter 5. But why would not only all those changes be accelerating but would also the rate of acceleration be speeding up from time to time?

NONLINEAR PRESSURES OF LIFE: OUR PEPPER PLANTS

How fast does life grow? And what may be salient resulting effects within the biosphere? Unlike inanimate nature, at least most of the time, life's growth has nonlinear characteristics as soon as the circumstances permit it. 'Nonlinear' means that when plotted on a graph, such processes do not show straight lines, but instead look like curves that are going up or down, in doing so exhibiting accelerating change. In a linear process, one week after sprouting a plant might have one single leave. After two weeks, it might have two leaves of the same size, after three weeks three similar leaves, and so on.

Yet that was not what our pepper plants were doing. After settling themselves in the soil, they began to grow rapidly, most visibly their leaves, both in numbers and in size, while also growing larger stems and roots. Such rapidly-increasing growth is known as an exponential, nonlinear, process. In this book, the term nonlinear growth is preferred, because in living nature, growth rates tend to vary, and, in consequence, do not always show neat exponential growth curves. Those varying growth rates very much depend on both the survival strategies of the species involved as well as on the circumstances they find themselves in.

As described in Chapter 2, I had noticed this nonlinear growth already in 2017. But it was only while again growing peppers in the spring of 2020 that I realized its general importance for the biosphere's history. I was then growing mostly Peruvian peppers, both *Aji amarillo* and *Rocoto*, the seeds of which had been ordered over the internet. Also a few *Capsicum annuum* and *Scotch Bonnet* seeds had been sown, which had come as complimentary gifts with those orders. All the plants, but especially the *Capsicum annuum*, suddenly began to grow very rapidly after having settled in the soil. The Peruvian peppers also showed growth spurts, but at considerably lower rates.

At that time, the COVID-19 virus pandemic was rapidly spreading around the world, which also exhibited a nonlinear growth pattern. I then realized that when the circumstances are sufficiently right, nonlinear growth is a general characteristic of life. Apparently, it occurs all over the food-capturing pyramid. In consequence, nonlinear growth needed to be included in my analysis of the biosphere's history.

Because I had missed the importance of nonlinear growth in 2017 as a general aspect of nature, none of that was measured at that time. But the many photos that were taken clearly show this process in action. A similar acceleration in growth can be seen on the 2020 photos, when it was not measured either. It then appeared so obvious to me that no need was felt to support this idea with specific measurements. But if one wanted to further explore this notion, such measurements need to be made. Doing so would constitute an entire research agenda.

Because the Sun was climbing higher in the sky during the spring while also the number of daylight hours rapidly increased, the available sunlight similarly increased, and so did the resulting ambient temperatures. All of this provided our plants with a rapidly-increasing supply of solar free energy. Furthermore, the springs of both 2017 and 2020 were unusually sunny in Amsterdam, especially during the afternoon when our plants were exposed to direct sunlight, which very much stimulated their growth spurts.

As a result, also the effects caused by our pepper plants on their immediately-surrounding biosphere exhibited nonlinear characteristics. Yet after the supply of sunlight began to decrease again in July and August, the plants' growth rates began to slow down as well. By the end of September, there was hardly any noticeable growth anymore of their leaves and stems. At the same time, the pepper pods kept growing while they had also increased in numbers. Also that process exhibited an initial nonlinear growth phase followed by a slowdown.

So far, we have only considered what happened to our pepper plants during one single season. But what if one takes into account the possible effects of our plants over a number of consecutive seasons, if they had been given the chance to survive instead of ending up in a Peruvian *ají* sauce? Because one single plant produces a great many seeds, here we see a second source of the pressure of life, which was already recognized by Darwin and Wallace, namely that life-forms tend to produce a great many progeny, more than will survive the struggle for existence. This capacity of life to produce large numbers of what is potentially the next generation is the second major nonlinear pressure exerted by life. As a result of this capacity, the numbers of most, if not all species, including humans, tend to grow in nonlinear ways as soon as the circumstances permit it.

In 2020, armed with this new awareness, I measured this second capacity of nonlinear growth for the *Capsicum annuum* plants. I had started out with only seven seeds. But those *Capsicum annuum* plants jointly produced seventeen fully grown pods. I counted the seeds of what appeared to be an average pod, and found that it contained 188 seeds. The total production of seeds by all those plants may therefore have been in the order of 3,200 seeds, which would yield about 460 seeds resulting from one single initial seed. Quite a nonlinear pressure of life!

And this is only the pressure exerted by one single generation of those pepper plants. Just imagine what would happen if all those seeds were fertile, while in every succeeding generation all of them got the chance to form new pepper plants. Starting with one single seed, in the seventh year there would be 9.5×10^{15} pepper plants. Assuming that, on average, they would be growing with a density of twenty plants per square meter, all those pepper plants combined would completely cover the entire planet, the landmasses as well as the oceans. And all of that after a mere seven years!

That is how strong the inbuilt capacity for nonlinear growth of our pepper plants is apparently honed into this shape by the processes of natural and human selection.

If those plants already contained so much pressure of life, as Vernadsky called it, what about the rest of life? And what about the resulting biospheric effects?

NONLINEAR GROWTH IN OTHER LIFE-FORMS

How would this work in other living organisms? First of all, the ways in which living species grow vary considerably. Most microorganisms including viruses do not to grow in size but instead in numbers. The reason for that is that they consist of only one single cell, or one single virus particle.

Cells are the basic building blocks of all life, viruses excepted. Microorganisms usually consist of only one single cell, while larger organisms including you, me, and our pepper plants consist of a great many cells. Such larger and more complex life-forms are therefore called multi-cellular organisms. The bodily growth of a multi-cellular individual is an aggregate effect of the growth in numbers of their cells. So, such individuals can growth both in body size and in numbers. Microorganisms and viruses, by contrast, can only grow in numbers but not in body size.

How would nonlinear growth affect multicellular organisms? The larger and more complex they are, the less conspicuous such effects may be for individuals, especially when their life-spans are much longer than one single growth season. Humans, for instance, tend to grow the fastest within the womb, within the first nine months of their existence, from one single fertilized egg cell to a fully-fledged human being. During the rest of our lifetime, such rapid nonlinear growth does not occur again. Many trees, by contrast, exhibit nonlinear growth spurts of their branches and leaves every spring, while human bodies do not do so. Yet the human numbers have also been growing nonlinearly, with ups and downs, most notably during the past 12,000 years. Today, this population growth appears to be slowing down again.

And, not least of all, also the emergence of new species often happens rather suddenly, so, in nonlinear ways, especially when the circumstances change or when life-forms invent novel ways of doing things, most notably perhaps more efficient ways of capturing free energy and matter. The nonlinear expansion of a range of novel species is known in biology as an adaptive radiation, usually resulting from biological innovations. It has long been known from the fossil record that such events tend to happen in spurts, which are followed by more stable periods. Most notably the US biologists Niles Eldredge (1943–) and Stephen J. Gould (1941–2002) drew attention to that phenomenon in their famous article ‘Punctuated equilibria: an alternative to phyletic gradualism’ (1972).

Such processes of sudden change are facilitated by the natural variation of cells, which is caused by changes in their molecules that carry hereditary information. As Darwin and Wallace already observed, this natural variation is subsequently filtered by the process of natural selection, in doing so allowing the sufficiently-adapted individuals to survive while the others perish. Yet when novel biological inventions happen or when the circumstances change, the criteria of natural selection change as well. This allows novel species to emerge very rapidly, so, in nonlinear ways, as long as they are sufficiently adapted to the prevailing circumstances.

In fact, the process of natural selection itself is possible thanks to the nonlinear growth of life. As Darwin observed, based on Malthus's famous essay, life-forms tend to produce more offspring than can possibly survive. This rapid growth in numbers is a direct result of the nonlinear growth potential of life. Coupled with the spontaneously-occurring variation among individuals, those large numbers make possible the selection for traits that help the fortunate individuals to survive and reproduce while the less fortunate are eliminated.

In sum: the growth of individuals and of their numbers, as well as the emergence of entire novel species, may all exhibit nonlinear growth. This makes one wonder how life's nonlinear growth is possible in the first place.

HOW IS THE NONLINEAR GROWTH OF LIFE POSSIBLE?

The basis for all of this nonlinear growth is that living cells can multiply nonlinearly as soon as the circumstances allow it. A cell usually multiplies by splitting into two virtually undistinguishable cells. So, one cell division produces two cells, the second generation four cells, the next one eight, and so on. This is typically a nonlinear process.

Some of those cells may die in the process, while others may be controlled in their growth by a great many causes. All of that may change their growth rates from exponential into less-rapidly evolving, but still nonlinear, processes, or perhaps even less than that. But unless there is a massive stagnation, or dying-off, of individual cells or complex bodies, such processes will remain nonlinear as long as the circumstances permit it, potentially leading to very rapid growth whenever and wherever possible. And because all living beings consist of cells, or, in the case of viruses, of even smaller and more basic survival units that can also reproduce themselves nonlinearly in large numbers, nonlinear growth is an inbuilt capacity of all life-forms.

This capacity of nonlinear growth by life is unique in the known universe. No inanimate processes known to us are doing that. It is the result of all life-forms actively seeking to capture sufficient free energy and matter, and, in doing so, seeking to increase their cells' numbers. As we saw before, this jointly produces the Fourth Law of Thermodynamics, but often in the form of nonlinear growth spurts. This process leads to more cells and individuals, to more competition for resources, and, as a result, to the emergence of novel and better-adapted species while others may go extinct.

The nonlinear processes of capturing free energy and matter can both increase and decrease very rapidly. They will increase as soon as species get the chance to do so, while they may rapidly decrease as soon as the circumstances deteriorate. A clear example of such a rapid decline is what happens at the end of our lives, as soon as we stop capturing free energy and matter and die as a result. But also the existence of entire species may suddenly come to an end. Perhaps the best-known example of such a situation is the demise of the land-dwelling dinosaurs after a meteorite hit our planet some sixty-six million years ago.

Yet in biological history, many species have exhibited growth and decline patterns that exhibit relatively little change over longer periods of time. Yet all such processes are dynamic steady-state regimes, with an inbuilt capacity for nonlinear change.

They continue to exist in such relatively stable ways as long as the circumstances remain similar. But they may change quite suddenly as soon as the circumstances change.

NONLINEAR PROCESSES IN INANIMATE NATURE

While considering the biosphere's history, we would need to consider not only life's capacity for nonlinear growth and decline, but also that of inanimate nature, in our case geology and cosmology. To what extent would such changes be gradual or nonlinear?

This issue of gradual or rapid geological change was already discussed in the nineteenth century between academics who saw Earth history as consisting of gradual continuous processes over long periods of time, including most notably perhaps the British naturalist Charles Lyell (1797–1875), and academics who emphasized the importance of rapid catastrophic events, led by the French naturalist Georges Cuvier (1769–1832). As so often in science, their academic positions may have been related to their personal experiences, in Cuvier's case: the French Revolution, while Lyell may have preferred the more gradual British societal changes. As has happened similarly often in science, both camps turned out to be correct, depending on the circumstances. Some processes happen gradually, while others occur suddenly.

Yet none of those natural inanimate processes exhibits forms of self-replication. Earthquakes and volcanic eruptions do not multiply, although they may trigger other earthquakes and volcanic eruptions in nearby regions that have become geologically unstable. But that is quite different from the growth of our pepper plants in spring, or of any other life form that finds itself in a nonlinear growth spurt.

The underlying reason for that difference is that all the cells of life actively harvest free energy and matter from outside while they can multiply as well, while earthquakes and volcanic eruptions do not have such capacities. But surely, also inanimate nature exhibits nonlinear aspects from time to time, such as rain showers, lightning strikes, inundations, droughts, rock and mud slides, meteorite impacts, supernova explosions, and volcanic eruptions, to mention a few of the more obvious sudden processes that have influenced our biosphere's history.

To my knowledge, the subject of nonlinear processes in nature has not yet systematically been analyzed in a general way. It may not yet even have been described as such. It may be so obvious and omnipresent that it is hard to see. In consequence, it may, so far, have escaped the academic attention.

WHAT ABOUT NONLINEAR EFFECTS WITHIN THE EVOLVING FOOD-CAPTURING PYRAMID?

Life's capacity of nonlinear growth must have produced a great many nonlinear effects within the food-capturing pyramid as it emerged and evolved over the course of time, and, in consequence, also in the biosphere's history. While the food-capturing pyramid grew and diversified, more, and more complex, interdependencies within it emerged, all with potential nonlinear growth and decline, as well as their resulting potentially nonlinear biospheric effects. Quite a dynamic food-capturing pyramid! Yet to the best of my knowledge, none of that has been described so far.

The COVID-19 pandemic spreading in 2020 offers an excellent example of such a nonlinear process. Much like how humans realized after a while that for countering this rapid viral growth, they needed to keep a certain physical distance to each other, especially within the tropics a similar situation inhibits the concentration of too many individuals of a certain species within one single area, because such a situation facilitates nonlinear predatory attacks by secondary captors.

Yet while local and regional populations may die out as a result of such nonlinear attacks, to my knowledge no entire species has gone extinct solely as a result of them, although that may have happened. Such species usually tend to bounce back and establish dynamic interdependency balances with their predators as part of their ever-continuing arms' races, which are all potentially nonlinear.

All those situations combined led to the emergence of very dynamic and ever-shifting balances of power and dependency among those species, and, in consequence, within the entire food-capturing pyramid. In consequence, this pyramid is a very dynamic regime with inbuilt tendencies for nonlinear expansion of all species as soon as they get the chance to do so. And because the conditions on Earth's surface vary, this has led to a widening range of increasingly-varied biospheric areas.

In sum: the nonlinear growth capacity of life has made the food-capturing pyramid very dynamic, because all the species involved can, in principle, engage in such activities. It also greatly adds to the pressure of life exerted by the Fourth Law of Thermodynamics: the tendency toward jointly maximizing the capture of the available free energy and matter, because it can make such effects occur very rapidly. It seems, therefore, fair to say that the Fourth Law of Thermodynamics exhibits an inherent tendency for nonlinear growth, which appears as soon as the circumstances are sufficiently right, in doing so leading to a great many rapidly-occurring and very complex processes of growth and decline.

HUMAN NONLINEAR GROWTH AND DECLINE

And what about nonlinear processes in human history? The short answer is that unlike any other species, humans have both a biological and cultural potential for nonlinear growth. As we will see in Chapter 7, while employing culture, human history can be described as the partial escape of one single species from the food-capturing pyramid. Our species subsequently began to make use of the rest of the biosphere in unprecedented ways, first of all by jointly seeking to maximize capturing all the available free energy and matter. This has led to unprecedented nonlinear human growth in concert with a nonlinear decline of a great many undomesticated species. Yet domesticated species, and sometimes also their predators, have fared well as long as humans have taken care of them.

Furthermore, humans are the only species that has exhibited intensive economic growth, sometimes in nonlinear ways, leading to unprecedented forms of material accumulation. By itself, such traits are not entirely absent among some animals, but humans have been doing this far more than any other species.

How has this been possible? Again, the answer is simple. Because ideas can be copied and spread cheaply and rapidly in nonlinear ways, cultural inventions and their effects often exhibit nonlinear characteristics, both in the ways they emerge and evolve and in their diffusion within and among societies. To be sure, the inventions do not reproduce themselves. No smartphone, for instance, has ever been observed to spontaneously split into two similar cell phones, and so on. Humans are responsible for their nonlinear growth by doing the copying as well as by transporting them all around the globe. Those processes are driven by worldwide competition and the resulting cultural selection and changes, including, most notably, the expectation of financial profits.

In other words, thanks to their learned cultural behavior, human societies have an additional inbuilt capacity for nonlinear growth. And because human societies can also exhibit biological nonlinear growth, our species has two different capacities for potential nonlinear growth, and decay: biological and cultural change. Nonlinear cultural decay may, for instance, occur during violent human encounters, most notably perhaps the destruction caused during wars, or as the effects of infectious diseases, such as what happened in the Americas after Europeans began contacting them systematically after 1492. But it may also happen by sheer negligence, of which the fire in 2019 of the Notre Dame basilica in Paris, France, offers a sad example.

Quite often, processes of nonlinear cultural growth and decay may happen simultaneously. The European conquest of the Americas, for instance, exhibited many examples of nonlinear decay of the indigenous populations as well as of a simultaneous European expansion. And so did the rise of personal computers in the 1980s, which very quickly ended the dominion of typewriters, including the great many secretaries that were operating them. Also in the rest of the living nature, processes of nonlinear growth and decay may happen simultaneously, especially when life-forms are feeding on other organisms.

WHERE DO WE GO FROM HERE?

As argued above, while considering the biosphere's history, we first of all need to look at all living organisms as captors of free energy and matter. That is their major goal because that is what keeps them going. Because life consists of matter, it also needs to harvest that. But to obtain matter and do something with it, life needs free energy.

By looking at the history of life and the biosphere in this way, we are no longer speaking of the emergence and evolution of certain species, but instead of the emergence and evolution of certain survival strategies of capturing free energy and matter, all in relation to each other, as part of the food-capturing web as a whole. To avoid any possible misunderstandings, I am not advocating here to stop looking at individuals or individual species. To the contrary, that is, and will remain, our principal empirical evidence. The argument here is, that for understanding the biosphere's history, it is necessary to examine all those species jointly in terms of survival strategies, together with all their mutual interdependencies and emergent effects.

The sociologist Norbert Elias argued that to understand both individual human beings and the societies they form with each other, we need to analyze them in terms of processes as interdependent people. For understanding the biosphere's history, we should do similarly for all living beings during all of their history, including all the emergent effects. That is not an entirely new point of view. Within the field of ecology, valiant attempts to do so have already been made. Yet to my knowledge, such analyses have not yet been extended to the entire history of life, let alone the biosphere's history. It is about time to start doing so, if we want to understand sufficiently well our current biosphere as well as our human position within it.

While pursuing such a goal, scientists may have been hindered by another bias, namely by the preference for examining the individuals of certain species, and certain species as a whole, while neglecting to investigate them systematically as part of the network of interdependencies that has connected them to all other species. This bias will be called Bias # 5 in this book. Much like in psychology, where individuals receive most of the attention without sufficiently examining the ways in which those persons are interconnected to other people, while calling such other people their 'social environment,' a similar bias may exist within the study of biological species.

To be sure, it is much easier to investigate examples of living species that can be observed in action. None of that can be studied, including their interdependencies with other species, while examining the remains of species that are no longer alive, which is the case for most species in the history of life. Paleontologists have expended great efforts on collecting the fossil remains of such species while seeking to interpret them. However valuable the resulting fossil record is, it is by necessity sparse and often individualistic in nature, also because quite often the bones of fossils are excavated while the surrounding soil is mostly, if not entirely discarded, in doing so potentially losing information about the ancient environment that this soil may contain.

Furthermore, a great many species may have lived on this planet without leaving any traces that we can currently recognize as such. This applies, in particular, to microorganisms, including viruses. Unless bounded together into larger structures such as bacterial mats, those little but omnipresent forms of life may have left hardly any traces in the fossil record, if at all, simply because they were so little compared to the more complex plants and animals. Yet they must have formed a most important portion of the food-capturing web as long as it has existed.

The fossil record is also sparse as a result of the processes of plate tectonics and erosion. Those processes tend to recycle our planet's surface, while also life contributes a great deal to such recycling processes. In fact, it has probably happened only rarely that certain individuals found themselves in a situation that favored fossilizing their remains. And even when that happened, over the course of time the biospheric recycling processes have destroyed most fossils, or buried them so deeply below our feet, or into mountains, that they cannot easily be found.

Only the fossils that have been accessible to us over the past 300 years or so have ended up in the fossil record, whenever and wherever scientists have gone to look for them. In addition, many fossils may have been picked up by curious human beings from earlier generations who did not know what they were, while they may have used them for other purposes such as religious rituals. Large-scale mining has also destroyed countless fossils, for instance those present in coal layers that have been exploited since the beginning of the industrial revolution, while most of them were subsequently burned. In doing so, humans destroyed forever any fossil evidence of ancient plants that such layers may have contained. A similar situation may pertain to petroleum reserves.

Yet however sparse and incomplete the current fossil record may be, it does not change the central idea proposed in this chapter that the history of life is first of all the history of the food-capturing web, including all its emergent effects. How could all of that lead to a description of the biosphere's history? And how would the humanity's history fit into such a scheme? Those questions will be discussed in the coming chapters.

BIBLIOGRAPHY

Alvarez, Walter. 2016. *A Most Improbable Journey: A Big History of Our Planet and Ourselves*. New York, W.W. Norton & Company.

Chaisson, Eric J. 2001. *Cosmic Evolution: The Rise of Complexity in Nature*. Cambridge, MA, Harvard University Press.

Corning, Peter A. 2005. *Holistic Darwinism: Synergy, Cybernetics, and the Bioeconomics of Evolution*. Chicago & London, University of Chicago Press.

Darwin, Charles. 1958. *The Autobiography of Charles Darwin, 1809–1882. With Original Omissions Restored, Edited with Appendix and Notes by His Grand-Daughter Nora Barlow*. London, St. James's Place, Collins.

Darwin, Charles. 1859. *On the Origin of Species by Means of Natural Selection, or the Preservation of Favoured Races in the Struggle for Life*. London, John Murray.

Darwin, Charles, Fitz-Roy, Robert, & King, Phillip Parker. 2015a. *Narrative of the Surveying Voyages of His Majesty's Ships Adventure and Beagle Between the Years 1826 and 1836.* Volume 2, Proceedings of the Second Expedition, 1831–1836. Cambridge, Cambridge University Press (1839).

Darwin, Charles, Fitz-Roy, Robert, & King, Phillip Parker. 2015b. *Narrative of the Surveying Voyages of His Majesty's Ships Adventure and Beagle between the Years 1826 and 1836.* Volume 3, Journal and Remarks 1832–1836. Cambridge, Cambridge University Press (1839).

Eldredge, Niles. & Gould, Stephen Jay. 1972. 'Punctuated equilibria: An alternative to phyletic gradualism'. In: Schopf, Thomas Joseph Morton (ed.) *Models in Paleobiology*. San Francisco, CA, Freeman, Cooper & Co (82–115). http://www.blackwellpublishing.com/ridley/classictexts/eldredge.pdf.

Humboldt, Alexander von. 2004. *Ansichten der Natur, mit wissenschaftlichen Erlaüterungen und sechs Farbtafeln, nach Skizzen des Autors*. Frankfurt am Main, Eichborn Verlag (1807).

McNeill, William H. 1976. *Plagues and Peoples*. Garden City, NY, Anchor Press/Doubleday.

Prigogine, Ilya & Stengers, Isabelle. 1984. *Order Out of Chaos: Man's New Dialogue with Nature*. London, Heinemann.

Raman, Aaswath P., Li, Wei, & Fan, Shanhui. 2019. 'Generating Light from Darkness.' *Joule* 3 (1–8).

Reeves, Hubert. 1991. *The Hour of Our Delight: Cosmic Evolution, Order, and Complexity*. New York, W. H. Freeman & Co.

Schrödinger, Erwin. 2013. *What is Life?* Cambridge, Cambridge University Press (1944).

Smil, Vaclav. 1999. *Energies: An Illustrated Guide to the Biosphere and Civilization*. Cambridge, MA, MIT Press.

Spier, Fred. 1996. *The Structure of Big History: From the Big Bang until Today*. Amsterdam, Amsterdam University Press.

Spier, Fred. 2011. 'Complexity in Big History.' *Cliodynamics* 2 (146–166). https://escholarship.org/uc/item/3tk971d2.

Spier, Fred. 2015. *Big History and the Future of Humanity, Second Edition*. Oxford, Wiley-Blackwell.

Spier, Fred. 2017. 'Complexity in big history.' In: David C. Krakauer, John Lewis Gaddis, & Kenneth Pomeranz (eds.) *History, Big History, Metahistory*. Santa Fe, NM, SFI (Santa Fe Institute) Press (205–233). (This is the same article as Spier 2011 in Cliodynamics).

Vernadsky, Vladimir I. 1998. *The Biosphere*. New York, Copernicus Springer Verlag (1926, Original Version in Russian).

4 What Needs to Be Considered While Writing our Biosphere's History?

Life is just one damned thing after another, whether it is private or public life.

British historian Arnold J. Toynbee (1889–1975) in his article 'You Can Pack Up Your Troubles' in the US monthly magazine *Woman's Home Companion* (April 1952, p. 4).

Heraclitus, I believe, says that all things pass and nothing stays, and comparing existing things to the flow of a river, he says you could not step twice into the same river.

The Greek philosopher Plato in: *Cratylus* 402a = A6, as quoted in the online *Stanford Encyclopedia of Philosophy* (accessed May 23, 2021).

The understanding of human societies, it seems to me, requires testable theoretical models which can help to determine and to explain the structure and direction of long-term social processes.

Norbert Elias, 'The Retreat of Sociologists into the Present,' in: Goudsblom & Mennell (eds.) (1998, p. 178).

I remember looking down from the crest of the highest Cordillera; the mind, undisturbed by minute details, was filled by the stupendous dimensions of the surrounding masses.

Charles Darwin, *Charles Darwin's Beagle Diary Edited by R.D. Keynes* (2001, p. 444).

INTRODUCTION

How can we write a history of the biosphere while using the ideas outlined in the previous chapters? That is a major challenge, not least because to the best of my knowledge, similar studies do not yet exist. In addition to Vernadsky's book *The Biosphere*, the major overviews of the biosphere's history known to me are Mikhail Budyko's *The Evolution of the Biosphere* (1986); James Lovelock's *Gaia: The Practical Science of Planetary Medicine* (1991) and *The Ages of Gaia: A Biography of Our Living Earth* (1995); Ian Bradbury's *The Biosphere* (1998); Vaclav Smil's *The Earth's Biosphere* (2002); Kenneth Hamblin & Eric Christiansen's *Earth's Dynamic Systems* (2004); as

DOI: 10.1201/9781003275350-4

well as a great many articles that outline partial aspects of the biosphere and its history, yet do not offer a novel synthesis of the biosphere's history along the lines suggested in this book. With the exception of Vernadsky's, Lovelock's and Smil's studies, none of the above-mentioned books was known to me while growing pepper plants. They were all found as a result of literature research performed during the spring of 2021.

Pioneering accounts of Earth history and/or the history of life that were already known to me include: Broda (1978), Cloud (1978,1988), Fox (1988), van Andel (1994), Emiliani (1995), MacDougal (1996), McSween Jr. (1997), Drury (1999), Lunine (1999), Niele (2005), and Smil (1991 and 1999). In most, if not all, these studies, a great deal of attention is paid to the role of life; the resulting energy flows through the trophic pyramid; as well as life's actions in transforming the biosphere. Similar pioneering studies about humanity's role within the biosphere with emphasis on its energy use include Odum (1971), Odum and Odum (1981), Cook (1976), Debeir, Deléage, and Hémery (1991), Smil (1994), de Vries and Goudsblom (2002), and Crosby (2006).

I may have missed something, and, if so, I request all readers to inform me. But none of those accounts proposes a history of the biosphere while employing the fresh theoretical aspects outlined in the previous chapters. What follows in the coming chapters should, therefore, be seen as a first preliminary proposal of our biosphere's history phrased in those novel terms.

Would it be possible at this stage to write a much longer, far more elaborate, and much more detailed overview of the biosphere's history? Such a challenge would require many more years of study as well as far more pages than this book can contain. Writing such an account is, therefore, a task for future generations. The main purpose of this book is, instead, first of all to outline those fresh principles and, in the second place, to use them for writing a first preliminary summary of our biosphere's history.

While doing so, we will inevitably encounter a great many gaps in our knowledge. A portion of that knowledge may already be available in the academic literature, while other portions may still need to be elucidated. As a result, while engaging in this enterprise, we will again face considerable ignorance. But perhaps not too much confusion, because we have now at our disposal a theoretical scaffolding that makes it possible, in principle, to summarize all of that in a coherent manner.

Why would we start considering the biosphere's history at its very beginning? A major advantage of doing so is, that all first beginnings tend to be simple. Today, the biosphere is staggeringly complex. Just think of how our pepper plants were transforming the soil in their flowerpot by growing roots, and by using those roots for moving water and minerals to places where they would otherwise not have been. And this represents only a first beginning of a summary of all the external effects caused by a few pepper plants growing in one single flowerpot. Obviously, the biosphere as a whole is far larger and far more complex.

Yet when life originated here on Earth, it must have been rather simple, while its first emergent effects must also have been rather limited. And early Earth itself was also simpler than it is today. Clearly, it is much easier to start a history at its first simple beginnings, while subsequently describing it as processes which, over the course of time, have led to far greater complexity, than seeking to describe today's hugely complex biosphere without understanding it as the result of a long historical process.

Because the first beginnings of life and its effects were tiny, they may have left very few traces, if any, that we can detect today. Would we expect to find traces of the first life-forms that inhabited our planet? The chances of doing so appear remote. The earliest evidence of such beginnings is probably long gone as a result of the biosphere's dynamic processes, such as erosion and plate tectonics, which have tended to destroy it. Yet in seeking to reconstruct our biosphere's history, such an early beginning is exactly what we need to consider.

In this account it is, therefore, assumed that the beginnings of such processes often happened earlier than the oldest traces of them that we can observe today. For instance, in this book, a starting time for the earliest life on our planet will be set at four billion years ago, and for the earliest human fire use at 1.5 million years ago. Those are inevitably rough estimates. But that does not matter too much here, because the aim of this book is not to establish precise starting times of such processes but rather to describe those processes as accurately as possible in general terms.

If we instead assumed that those processes began only at the time when they left the first clear evidence at our disposal today, as is often done, especially in deep time, we would miss those crucial first beginnings, because they must have happened earlier than that. In consequence, we might not understand such processes sufficiently well. The bias of equating the period of which we have the earliest available evidence of novel processes with their 'real' beginnings will be called Bias # 6 in this book.

The description and analysis of our biosphere's history will not contain any novel empirical data. The emphasis here will be on the reinterpretation of those data by using the fresh theoretical principles that were discovered while growing pepper plants.[1]

WHAT ABOUT CHRONOLOGY AND PHASEOLOGY?

"Life is just one damned thing after another, whether it is private or public life." In this famous sentence, the British historian Arnold J. Toynbee (1889–1975) summarized his view that history consists of events that follow each other. By saying that "one cannot step twice into the same river," a similar, but more precise, insight was already expressed some 2,500 years earlier by the ancient Greek philosopher Heraclitus, who lived around 500 BCE in the town of Ephesus, situated in what today is western Turkey. Both quotations point at history as consisting of processes. Yet the issue of how to accurately describe historical change in terms of processes has not yet sufficiently been resolved.

While seeking to do that, first a chronological order needs to be established. As Norbert Elias explored in his pioneering book *Time: An Essay* (1992), during human history our species has devised increasingly precise ways of measuring ongoing processes by comparing them with the duration of natural cycles that appeared sufficiently regular, such as the rhythm of day and night; of the Moon's trajectory along the sky; and of the annual sequence of the seasons. That is why we still count history in terms of days, months, and years, even though today, precise time keeping is determined by measuring the far more regular cycles of atomic processes.

[1] Earlier versions of this chapter included a discussion of the fundamental possibilities and limitations of writing histories. But because this argument might interrupt the flow of this book's central argument too much, I decided not to include it, and save it instead for another occasion, however relevant this subject may be for writing a biosphere's history.

The determination and measurement of long-term processes, such as our biosphere's history, while placing shorter historical processes within such a time line, is known as chronology. Such time lines form the primary backbone of all academic historical studies, large and small.

But is that all that needs to be done? Is history indeed no more than just one damned thing after another, whether it is private or public life? How would we, for instance, determine which aspects are important to mention, and which are not, or less important, given our aims and the related theoretical perspectives? This issue appears urgent, because if we could not make such judgments, we would need to include everything we know within our historical accounts. And doing so would inevitably lead to impossibly long and detailed accounts that no one would want to pay attention to.

In other words, is it possible to discern larger patterns in history that help elucidate it, while not mentioning all the details? If so, how would we do that? For human history, Norbert Elias had an answer to that question, namely the sentence quoted at the beginning of this chapter:

> The understanding of human societies, it seems to me, requires testable theoretical models which can help to determine and to explain the structure and direction of long-term social processes.

So, in his view, we need to investigate long-term processes and see to what extent they exhibit certain patterns, which should be explained by using testable models.

How would we find social patterns? In doing so, we make use of terms that already exist in daily language, such as families, schools, universities, businesses, religious organizations, states, etc. All those terms indicate constellations of interdependent people that can be interpreted as ongoing processes. The next step is to determine how those social constellations have changed over time, and why. Can we, for instance, discern periods of relative stability followed by rapid change? Or was everything in continuous change? Or do we perhaps observe certain oscillations? Those examples offer only a few possibilities of such changes.

The changes of social processes can be delineated in terms of phases. For human history as a whole, for instance, we can say that for most of its existence, people lived as gatherers and hunters. Yet a new phase began when some people began to practice agriculture and animal husbandry, in doing so slowly but surely replacing the earlier gatherers and hunters. More recently, people began to make a living by using fossil fuels for productive purposes. That signaled the beginning of another novel phase in human history known as the industrial revolution. Much like agriculture and animal husbandry, ever since the industrial mode of production emerged, it has deeply influenced all the earlier social processes. Today, the information revolution is having a similar effect.

This short overview provides, of course, only a very general scheme of human history as a process. A slightly more refined outline would include the emergence and development of states, including their increasing domination of the earlier lifestyles. It would also mention the first wave of globalization that took place after Columbus and his crew unwittingly stepped ashore on what later would be called the Americas. In line with such an approach, the industrial revolution could be seen as the second wave of globalization, while the current information revolution would be the third such a wave. Such a process approach of human history was adopted in my books *The Structure of Big History* (1996) and *Big History and the Future of Humanity* (2010, 2015).

Within these general processes, more processes and their phases can be distinguished. Could we think, for instance, of an incipient phase of agriculture, followed by a phase of its rapid expansion and diversification, which, in its turn, was followed by period characterized by the intensification of domestication, including increasing long-distance exchanges of domesticates? Examining all of that in more detail would require studying large amounts of empirical evidence as well as a great many earlier studies, all of which could subsequently be summarized within such a scheme. Furthermore, within such large processes, a great many smaller processes could be distinguished, including, for instance, personal, family, and village histories. And all those processes could, in their turn, be seen as part of even larger cosmic processes, all the way up to big history as a whole, all in terms of interlinked processes.

As explained in Chapter 1, while studying cultural anthropology and sociology in the 1980s, thanks to Norbert Elias's sociology I learned to think in terms of social processes and process models, which has helped me a great deal to understand reality and its history in better ways. I was first stimulated to do so by Mart Bax and later also by Johan Goudsblom. In his illuminating essay *Human History and Long-Term Processes; Toward a Synthesis of Chronology and Phaseology* (1996), Goudsblom explained those issues very clearly.

Mart Bax in his office at the Free University Amsterdam, spring of 1985. (Photo by the author)

Johan Goudsblom lecturing during the One Day Conference on Civilizational Processes, Amsterdam, January 14, 1994. (Photo by the author)

A great deal more can be said about history as consisting of processes. But that would fall outside the scope of this book, which focuses on the biosphere's history. So, in our case, the question to be addressed is: which important processes and phases can be delineated in the biosphere's history?

WHAT ABOUT THE PHASES OF OUR BIOSPHERE'S HISTORY?

How can the phases of the biosphere's history be delineated? Because we are dealing with a complex planetary situation, this question requires a complex answer. Yet by examining all the free energy flows that have powered our biosphere during its entire history, we can discern some simplicity in it.

As mentioned before, Earth has two major sources of free energy: solar energy from outside, and geothermal energy from inside. Those sources are independent of each other. Solar free energy is a byproduct of the nuclear fusion reactions taking place deep inside the Sun, while geothermal free energy results from nuclear fission reactions deep within Earth. Those fission reactions consist of heavy chemical elements that spontaneously fall apart because they are unstable, which emits free energy in the form of heat and radiation.

Those flows of solar and geothermal free energy would have started almost simultaneously as soon as the Sun and the Earth had emerged. Yet they have their own independent evolutionary pathways. The fusion reactions that take place inside the Sun do not influence the nuclear breakdown of chemical elements deep within the Earth, or vice versa. Those two independent free energy sources have jointly powered the biosphere ever since it emerged.

The situation on Earth is, in fact, even more complex, because there are also sources of cosmic free energy and matter that reach our planet. Some of those flows arrive from inside the Solar System in the form of asteroid impacts, and also, for instance, through the gravitational effects of the Sun and the Moon, which cause the tides, as well as because the giant planet Jupiter keeps pulling at Earth's orbit. In doing so, it changes it in regular ways. And some cosmic effects may come from even farther away, such as the ones caused by nearby exploding large stars called supernovae.

All those cosmic processes have their own relatively-independent pathways and phases. Yet little or none of that cosmic free energy is captured by life, because most such flows occur suddenly while they tend to be short-lived. Life cannot easily capture such free energy flows, because its cells are delicate. They consist of a great many complex chemical processes that usually operate at moderate temperatures. Living cells need, therefore, similarly moderate free energy flows for their sustenance, such as the ones provided by geothermal and solar energy. But they cannot harvest sudden, short, and sometimes powerful, free energy flows.

As a result, the biosphere's history has been influenced by at least three major independent sources of free energy: geothermal, solar, and cosmic free energy. All those flows exhibit change over time, while they are all characterized by their own particular phases. Yet within the biosphere, all those effects have become intertwined within our extraordinarily complex biosphere.

Yet there is one more major source of complexity within the biosphere, namely all the effects caused by life, the history of which exhibits its own particular phases. While those phases of life have very much depended on the phases of the solar and geothermal free energy flows, life has also evolved a number of relatively-independent phases of its own, most notably the emergence of novel species, their changing interdependencies, and their external effects.

As a result of this situation, the biosphere's history consists of a very complex set of processes, with increasing numbers of feedback loops over the course of time, all of which need to be considered jointly and systematically. How can we devise a scheme in terms of processes with specific phases that does justice to this extraordinarily complex situation? That is the challenge we are facing while seeking to describe the biosphere's history.

My solution to this issue is as follows. However important geological, solar, and cosmic influences have been for our biosphere's history, they have not turned our planet's outer layer into a biosphere all by themselves. Instead, they have facilitated and conditioned the opportunities and limitations for life to do so. In the final analysis, living beings and their effects have jointly turned our planet's outer shell into a biosphere, through relatively-autonomous processes that were not completely determined by geothermal, solar, and cosmic free energy flows.

As an example, we could think of how our pepper and avocado plants grew. Their situation was conditioned by the circumstances within our apartment, such as the flowerpot; the ambient temperature; the quality of the soil; as well as the amounts of sunshine, water, and fertilizer that they received. Yet our pepper and avocado plants did not respond to those rather similar conditions in similar ways. Instead, they exhibited their own particular survival strategies. As a result, the character, and phases, of their little histories as described in Chapter 2 cannot completely be explained by those living conditions, even though they were important.

The same applies to life's history as a whole. Formulated in process terms, life's history exhibits a level of greater complexity and relative autonomy that cannot completely be reduced to, and explained by, geothermal, solar, and cosmic free energy flows, however important these have been in facilitating and conditioning the history of life.

In consequence, the phases of life's history appear to offer the most straightforward solution to the question of how to structure the biosphere's history. In our account, they will serve as its principal theoretical backbone, while the phases of geothermal, solar, and cosmic free energy flows will play a secondary, but still very important, role.

This choice raises the following questions: which phases of life could serve such a purpose, and how could they be determined? In this book, the choice was made to use the principal phases of life's survival strategies concerning the capture of free energy, while also paying due attention to the changing cosmic, solar, and geological influences. This stance will be further elaborated in the Chapters 5 and 7.

As a result of all those complications, the analysis of our biosphere's history presented in this book should be considered what natural scientists call 'a first-order approach:' a first preliminary model of reality. Such models are very common within the natural sciences as a first step in seeking to understand complex reality and its past.

WHAT ABOUT CHANCE AND NECESSITY IN THE BIOSPHERE'S HISTORY?

What are the roles of chance and necessity in the biosphere's history? In the last chapter of his book *A Most Improbable Journey: A Big History of Our Planet and Ourselves* (2016), the US geologist Walter Alvarez (1940–) argued that because everything in history so much depends on chance and contingency, the current situation on our planet is most improbable. I could not agree more with such a view. Yet the biosphere's history is not totally chaotic. To the contrary, it exhibits clearly discernable processes with similarly clear phases. How is this possible, if indeed there is so much chance and contingency in that history?

Here, we are dealing with the general problem of chance and necessity in history, which has been discussed for over two millennia. At the beginning of Chapter 2 of my book *Big History and the Future of Humanity* (2010, 2015) my position on these questions was stated as follows:

> explaining the past always implies striking a balance between chance and necessity. This point of view had already been expressed by the natural philosopher Democritus of ancient Greece (460–370 BCE), while more recently French biochemist Jacques Monod (1910–1976 CE) said essentially the same (with proper reference to Democritus).

> My explanatory scheme is about necessity. It consists of general trends that not only make possible certain situations but also constrain them. Yet within these boundaries there is ample room for chance.

If so, what is constraining chance effects in the biosphere's history, and, in doing so, produces processes with clear phases? In my weblog of 2017 titled *How Can We Examine Chance and Necessity in Big History?* this issue was compared with throwing a die. The outcome of every single throw cannot precisely be predicted. But because our die consists of sides that are numbered one to six, it is possible to forecast with absolute certainty that numbers lower than one and higher than six will not occur. So, in this case the available numbers are constraining the outcomes. But within those limits, a great deal of chance is possible.

A similar situation pertains to the biosphere's history. Those constraints include the size of our planet; its position within the Solar System and all the resulting effects; the process of Earth's formation as part of a long cosmic history, which provided the chemical elements needed for our planet's current complexity. All of that, and much more, has both facilitated and constrained our biosphere's history. In doing so, those conditions made possible processes with certain structures and phases. Yet within all of that, an enormous amount of chance has happened as well.

While studying human history in the 1980s and 1990s, the Dutch sociologist Johan Goudsblom and myself agreed on the formulation that we were not trying to explain how history *had* to happen, but that instead we sought to explain how it *could have* happened. We had reached that conclusion independently because of the role that chance and contingency play in history.

The same also applies to the biosphere's history. Clearly, it could have gone differently in many ways. Yet many of its patterns, which will be described in the Chapters 5 and 7, show a remarkable internal consistency and direction, even though no one has been planning them. For instance, in the beginning, our biosphere was very simple, while subsequently it became increasingly complex and varied. It happened this way, and not the other way around, or in totally chaotic fluctuations. It is therefore not a history without any clearly discernable patterns.

This situation very much resembles the process of plate tectonics, which has shaped the configuration of the oceans and continents over billions of years. Because too many chance effects are involved, the precise form and position of the continents cannot be predicted at any given moment. But over the course of time, this process has produced growing continents and decreasing oceans. Similarly, after life had emerged on our planet, its increasing biodiversity was almost inevitable, cosmic disasters excepted, given Earth's chemical starting composition and its position within the Solar System. That is how structures and phases emerge out of what appear, at first sight, to be pure chance effects.

HOW CAN WE DEAL WITH DETAILS AND OVERVIEWS IN THE BIOSPHERE'S HISTORY?

How can we deal with both details and overviews in history accounts? In seeking to answer this question, let us first consider what humans do while watching photos and especially movies. In doing so, our eyes are able to focus on both the details and the overviews, but not on both at the same time. This can happen thanks to our brain's capacity of mentally zooming in and out, much like the zoom lenses of cell phones and digital cameras.

While mentally zooming in, the details appear much larger than while watching the overview. This mental capability can also be observed, for instance while driving a car, especially when something of interest attracts the attention. Apparently, there has been an evolutionary advantage for developing such a mental capability, which, as always, must have been honed by natural selection.

So, while watching, people are zooming in at certain moments to see the details followed by zooming out to see overviews, while switching between those two modes usually happens almost instantaneously. Hollywood movie producers make good use of this capability, by continuously zooming in and out. As a result, while watching movies we are confronted by two zooming activities: one provided by the movie maker, and the other one that takes place inside our brains.

Not only eyes and brains jointly perform such zooming activities but also ears and brains can do so, by suddenly focusing on specific sounds. Yet in spoken and written language, we cannot produce sounds and sentences that can similarly quickly switch between details and overviews. Doing so takes a much longer and more strenuous mental effort. Would this help to explain why people have been putting images in books as soon as that was feasible, already many centuries before printing was invented, while today's readers interested in a new book often start by leafing through the pages in search of pictures?

This raises the question why some books contain a great many pictures while others do not, or to a far lesser extent. A similar question can be posed for academic lectures. An important reason why academic books tend to contain relatively few pictures is that those books are often printed in small numbers because the expected sales volumes tend to be limited. And because including pictures costs money, academic publishers tend to limit the number of pictures that are allowed to be included. In books with larger expected sales volumes, this is less of an issue, which helps to explain why such productions tend to contain more images.

Yet the importance of pictures for conveying information remains undeniable. Already Alexander von Humboldt was well aware of this situation. As he formulated it: 'Even when one reads the best descriptions of the Andes Cordillera, the Swiss Alps, the Caucasus, or the sunken Trap Chain of the South Sea, these readings will never evoke the ideas offered by the view of a geognostic map.'[2]

In sum: pictures and movies can provide accounts of reality in ways that spoken or written accounts cannot, or only to a lesser extent. Furthermore, visual images can connect more directly to the brain's accumulated store of mental images than those created by words, because the latter first need to be translated into mental images before they can connect to the brain's already-existing mental images.

---

2 This is my translation into English of Alexander von Humboldt's text in French from his book *Essay de Pasigraphie* (1803) quoted in: Erdmann & Lubrich (2019), p. 22.

But while visual images of reality are easier and quicker to access, they cannot offer interpretations of it. Doing so is a major strength of language. This immediately explains why documentaries always contain narrators, or written texts that explain what is going on, as well as why photographs and movies without explanations are often hard to interpret.

So, how would we deal in the biosphere's history with details and overviews? To me and many others, the best solution consists of first providing the overviews and subsequently some details that illustrate those larger trends. Doing so may connect the best to our brain's capacity for zooming in and out, while it costs the least brain power for placing details within a larger mental context.

How many details would need to be included to reach a good understanding of what is trying to be conveyed? This very much depends on the audience. Some people like a great many details, while others prefer overviews. In this respect, it will be impossible to satisfy everybody. But whatever choices are made, it remains important that any details that may contradict the overview are mentioned. In fact, the author must think very hard in advance to make sure that such contradictions do not occur. Doing so that may be the hardest part of writing an academic text.

And not least of all, the publisher usually puts a limit on the book's size, and, in consequence, on the number of details that can be included. With the exception of fiction and textbooks, in our days hardly anybody reads long books anymore, perhaps also because of the prevalence of a great many easily-accessible photos and movies. It therefore makes perfect sense to limit the size of a non-fiction book.

While considering all these issues, I realized that I was facing a dilemma not unlike the one faced by a shipping company that transports oil from the Middle East to Europe. If those so-called very large crude carriers are too big, they will not be able to pass through the Suez Canal, which offers the shortest possible sea route that connects those two areas. Instead, such a fully-loaded tanker would have to sail around Africa, which involves a considerably longer distance.

A similarly longer excursion of our biosphere's history overloaded with details does not appear to be advisable within the context of this book, the first aim of which is to present a fresh theoretical view as well as its first preliminary application. In the Chapters 5 and 7, I have therefore opted for the Suez Canal variant: fewer details and a resulting shorter account. In fact, the huge container ships that are currently able to pass that famous canal already transport enormous cargos. I hope not to have overloaded even the Suez Canal variant of the biosphere's history in the coming chapters with too many details.

This brings up another issue, namely: the larger overviews are, the harder it becomes to reference them properly, because doing so exhaustively would require a considerable number of references after almost every sentence. While that may work well in scientific articles, such a writing style does not produce very readable books. In this account, the preferred option for dealing with this issue is, therefore, to mention the principal sources in the bibliography without further referencing them in the main text. It is not an entirely satisfactory solution. But it is the best one I can think of right now. In that respect, the reader is requested to keep in mind that the main aim of this book is to reinterpret the existing knowledge of the biosphere's history in terms of a fresh theoretical approach.

ONE MORE BIAS: ACCOUNTS OF LIFE'S HISTORY TEND TO FOCUS ON COMPLEX ORGANISMS

As part of our reflections, we need to consider another common bias, namely the often-encountered idea that the major trend in biological evolution would be the emergence and development of increasingly-complex life-forms, culminating with humans. Even though humans are indeed the most complex species that has ever lived on this planet, such a view is only partially correct, because it neglects the history of the food-capturing web as a whole. A more correct view would be to state that over the course of time, with ups and downs, the food-capturing pyramid as a whole has become increasingly complex.

The bias that the main trend in biological evolution would be the one toward more complex species will be called Bias # 7. It is different from Bias # 4: the tendency to focus the attention to increasingly complex species, although it may result from it. In Bias # 7, Bias # 4 is projected onto the entire history of life, openly, or implicitly, declaring it to be its major trend, which it is not. The Biases # 4 and # 7 have had profound consequences for interpreting the history of life.

As a result of those two biases there is, for instance, more money available for researching bigger and more spectacular fossils, especially if they can be considered precursors of humans or otherwise capture the imagination of the masses. What would be a bigger moneymaker, would you think, a movie about dinosaurs; one about the Jurassic vegetation eaten by dinosaurs; or one about the microorganisms that lived during that period? This social bias has a direct consequence for academic funding, which, in consequence, tends to focus the attention on more complex species.

Such a bias can not only be observed in paleontology but also within the humanities. For instance, about twenty years ago, a Dutch scholar of ancient Greece told me that it was virtually impossible to obtain a research grant for excavating ordinary houses dating from that period. Instead, virtually all the money was spent on excavating more spectacular buildings such as temples. As a result, he said, we still hardly know what such ordinary dwellings looked like.

There is no reason to suspect that this situation has changed significantly since that time. I witnessed similar situations also elsewhere in the world, including the ancient Mayas in Yucatan and the Incas in and around Cusco, Peru. All of this results from the bias toward larger secondary and tertiary captors of free energy, including the spectacular cultural human achievements, while the great many primary captors of free energy receive far less attention, especially when they are little, even though they capture all the free energy that keeps the entire the food-capturing pyramid going, including making all those spectacular and prestigious things.

What follows in the next chapters represents my best effort at reorganizing the known academic furniture in the room with the aid of our fresh insights to achieve a novel synthesis that is hopefully a little more 'reality-congruent,' as Norbert Elias would have called it. Let us therefore now move on to the next chapter and begin examining the biosphere's history in terms of what we have learned from growing peppers.

BIBLIOGRAPHY

Andel, Tjeerd Hendrik van. 1994. *New Views on an Old Planet: A History of Global Change.* Cambridge, Cambridge University Press (1985).

Alvarez, Walter. 2016. *A Most Improbable Journey: A Big History of Our Planet and Ourselves.* New York, W.W. Norton & Company.

Bradbury Ian K. 1998. *The Biosphere*, Second Edition. Chichester, John Wiley & Sons.

Broda, E. 1978. *The Evolution of Bioenergetic Processes (revised reprint).* Oxford & New York, Pergamon Press.

Budyko, M.I. 1986. *The Evolution of the Biosphere.* Dordrecht & Holland, D. Reidel Publishing Company (original edition in Russian 1984).

Cloud, Preston. 1978. *Cosmos, Earth and Man: A Short History of the Universe.* New Haven & London, Yale University Press.

Cloud, Preston. 1988. *Oasis in Space: Earth History from the Beginning.* New York & London, W.W. Norton & Company.

Cook, Earl. 1976. *Man, Energy, Society.* San Francisco, CA, W.H. Freeman & Co.

Crosby, Alfred W. 2006. *Children of the Sun: A History of Humanity's Unappeasable Appetite for Energy.* New York, W. W. Norton & Co.

Darwin, Charles. 1859. *On the Origin of Species by Means of Natural Selection, or the Preservation of Favoured Races in the Struggle for Life.* London, John Murray.

Darwin, Charles. 2001. *Charles Darwin's Beagle Diary Edited by R.D. Keynes.* Cambridge, Cambridge University Press (original manuscript 1836).

Debeir, Jean-Claude, Deléage, Jean-Paul & Hémery, Daniel. 1991. *In the Servitude of Power: Energy and Civilization through the Ages.* London & Atlantic Highlands, NJ, Zed Books (1986).

Drury, Stephen. 1999. *Stepping Stones: The Making of Our Home World.* Oxford, Oxford University Press.

Elias, Norbert. 1992. *Time: An Essay.* Oxford, Blackwell.

Emiliani, Cesare. 1995. *Planet Earth: Cosmology, Geology, and the Evolution of Life and Environment.* Cambridge, Cambridge University Press.

Erdmann, Dominik & Lubrich, Oliver (eds.). 2019. *Alexander von Humboldt: Das zeichnerische Werk.* Darmstadt, WBG (Wissenschaftliche Buchgesellschaft).

Fox, Ronald F. 1988. E*nergy and the Evolution of Life.* New York, W. H. Freeman & Co.

Goudsblom, Johan. 1996. 'Human history and long-term processes; Toward a synthesis of chronology and phaseology.' In: Goudsblom, Johan, Jones, Eric & Mennell, Stephen (eds.) *The Course of Human History: Economic Growth, Social Process, and Civilization.* New York, Armond & London, M.E. Sharpe (15–30).

Goudsblom, Johan & Mennell, Stephen (eds.). 1998. *The Norbert Elias Reader.* Oxford, Blackwell Publishers.

Hamblin, Kenneth W. & Christiansen, Eric H. 2004. *Earth's Dynamic Systems*, Tenth Edition. Upper Saddle River, NJ, Pearson Education Inc. / Prentice-Hall, Inc.

Humboldt, Alexander von. 1845. *Kosmos: Entwurf einer physischen Weltbeschreibung.* Stuttgart & Augsburg, J. G. Cotta'scher Verlag (1845–62).

Humboldt, Alexander von. 1997. *Cosmos*, vol. 1, Foundations of Natural History. Baltimore, MD, Johns Hopkins University Press (1858).

Humboldt, Alexander von. 2004. *Ansichten der Natur, mit wissenschaftlichen Erlaüterungen und sechs Farbtafeln, nach Skizzen des Autors.* Frankfurt am Main, Eichborn Verlag (1807).

Lunine, Jonathan I. 1999. *Earth: Evolution of a Habitable World.* Cambridge, Cambridge University Press.

Lovelock, James E. 1987. *Gaia: A New Look at Life on Earth.* Oxford & New York, Oxford University Press (1979).

Lovelock, James. 1995. *The Ages of Gaia: A Biography of Our Living Earth*. New York, W. W. Norton & Company.
Lovelock, James. 2000. *Gaia: The Practical Science of Planetary Medicine*. Oxford & New York, Oxford University Press (1991).
MacDougal, J. Doug 1996. *A Short History of Planet Earth: Mountains, Mammals, Ice, and Fire*. New York, John Wiley & Sons.
Margulis, Lynn & Sagan, Dorion. 1995. *What is Life? (Foreword by Niles Eldredge)*. London, Weidenfeld & Nicholson.
Margulis, Lynn & Sagan, Dorion. 1997. *Microcosmos: Four Billion Years of Microbial Evolution*. Berkeley, CA, University of California Press (1986).
McNeill, William H. 1986. *Mythistory and Other Essays*. Chicago & London, University of Chicago Press.
McSween Jr., Harry Y. 1997. *Fanfare for Earth: The Origin of Our Planet and Life*. New York, St. Martin' s Press.
Monod, Jacques. 1971. *Chance and Necessity*. New York, Alfred Knopf.
Niele, Frank. 2005. *Energy: Engine of Evolution*. Amsterdam, Elsevier/Shell Global Solutions.
Odum, Howard T. 1971. *Environment, Power and Society*. New York, John Wiley & Sons, Inc.
Odum, Howard T. & Odum, Elizabeth C. 1981. *Energy Basis for Man and Nature*, Second Edition. New York, McGraw-Hill Book Company.
Smil, Vaclav. 1991. *General Energetics: Energy in the Biosphere and Civilization*. New York, John Wiley & Sons.
Smil, Vaclav. 1994. *Energy in World History*. Boulder, CO, Westview Press.
Smil, Vaclav. 1999. *Energies: An Illustrated Guide to the Biosphere and Civilization*. Cambridge, MA, MIT Press.
Smil, Vaclav. 2002. *The Earth's Biosphere: Evolution, Dynamics, and Change*. Cambridge, MA, MIT Press.
Spier, Fred. 1994. *Religious Regimes in Peru: Religion and State Development in a Long-Term Perspective and the Effects in the Andean Village of Zurite*. Amsterdam, Amsterdam University Press. Downloadable for free as a pdf at: http://www.bighistory.info/bhi_005_035.htm.
Spier, Fred. 1995. *San Nicolás de Zurite: Religion and Daily Life of an Andean Village in a Changing World*. Amsterdam, VU University Press. Downloadable for free as a pdf at: http://www.bighistory.info/bhi_005_035.htm.
Spier, Fred. 1996. *The Structure of Big History: From the Big Bang until Today*. Amsterdam, Amsterdam University Press.
Spier, Fred. 2010. *Big History and the Future of Humanity*. Oxford, Wiley-Blackwell.
Spier, Fred. 2015. *Big History and the Future of Humanity, Second Edition*. Oxford, Wiley-Blackwell.
Spier, Fred. 2017. *How can we examine chance and necessity in big history?* March 2, http://www.bighistory.info/bhi_005_029.htm.
Toynbee, Arnold J. 1934–1961. *A Study of History*, 12 vols. Oxford, Oxford University Press.
Toynbee, Arnold J. 1952. 'You can pack up your troubles.' *Woman's Home Companion*, April 1952 (4–6).
Vernadsky, Vladimir I. 1998. *The Biosphere*. New York, Copernicus Springer Verlag (1926, Original Version in Russian).
Vries, Bert de & Goudsblom, Johan. 2002. *Mappae Mundi: Humans and their Habitats in a Long-Term Socio-Ecological Perspective, Myths, Maps and Models*. Amsterdam, Amsterdam University Press.

5 A Fresh History of the Biosphere before Humans

> Beginning with the depths of space and the regions of remotest nebulae, we will gradually descend through the starry zone to which our solar system belongs, to our own terrestrial spheroid, circled by air and ocean, there to direct our attention to its form, temperature, and magnetic tension, and to consider the fullness of organic life unfolding itself upon its surface beneath the vivifying influence of light.
>
> Alexander von Humboldt, *Cosmos*, vol. 1 (1997, pp. 79–80), original German edition (1845, p. 80).

> The outer layer of the Earth must, therefore, not be considered as a region of matter alone but also as a region of energy and a source of transformation of the planet.
>
> Vladimir Vernadsky, *The Biosphere* (1998, p. 44), original Russian edition (1926).

> Living things profoundly influence and are influenced by Earth's physical processes. They have played a major role in shaping and recording planetary history from time immemorial. The mutual feedbacks between life and its physical surroundings – biospheric and biogeologic processes – can never be far from our thoughts as we attempt to reconstruct that history.
>
> Preston Cloud, *Oasis in Space: Earth History from the Beginning* (1988, p. 42).

> No one ought to feel surprise at much remaining as yet unexplained in regard to the origin of species and varieties, if he makes due allowance for our profound ignorance in regard to the mutual relations of all the beings which live around us. [...] Still less do we know of the mutual relations of the numerous inhabitants of this world during the many past geological epochs of its history.
>
> Charles Darwin, *On the Origin of Species* (1859, Introduction, p. 6).

INTRODUCTION: GLOBOSPHERES AND BIOSPHERES

The Earth is unique in the known universe because it contains life that has shaped its biosphere over billions of years. Yet its deeper interior, the largest portion of our home planet, is considerably less unique. In fact, the deeper layers of our sister planet Venus may be rather similar, even though it does not have plate tectonics or a biosphere. However, the focus of our account is on the history of Earth's unique outer shell, its biosphere, where life and its effects can be found.

DOI: 10.1201/9781003275350-5

All the other planets of our Solar System and many moons also have outer shells, many of which exhibit considerable complexity. But as far as we know, none of those outer shells contains life. In an analogy with Earth's biosphere, in our account such lifeless yet often complex planetary and lunar outer shells will be called 'globospheres,' which literally means 'spheres surrounding globes.' This may sound like an unusual term. But I became quickly accustomed to it, because it turned out to be useful. Much like the term 'biosphere,' also the word 'globosphere' is a mental concept that summarizes a large amount of empirical data, while both terms indicate processes that change over time.

How could, for instance, the current scientific investigations into the planet Mars' outer shell be summarized using this term? A great deal of recent research on the red planet is being performed by remotely-controlled spacecraft: stationary landers, rovers, and orbiters. And a considerable portion of that scientific work is aimed at trying to find aspects within Mars' globosphere that cannot be explained by the dynamics of lifeless globospheres: most notably, traces that indicate whether the red planet once sported life, or whether it perhaps still does, and, in consequence, whether Mars once may have had a biosphere instead of only a lifeless globosphere.

Clear empirical evidence has been found indicating that billions of years ago, Mars had a much thicker atmosphere than today, as well as considerable amounts of liquid or frozen water on its surface. But that does not necessarily indicate the presence of life. Also lifeless globospheres can evolve such complex landscapes. As yet, no clear traces of life have been found on Mars, the ambiguous results of the Viking landers in the 1970s perhaps excepted. But those results are still considered controversial.

In other words, within Mars' globosphere, scientists are looking for traces of its potential current or former biosphere. As part of that, one would expect a lifeless globosphere with a biospheric past to be different from one that did not have such a history. If evidence were found indicating a past biosphere that no longer exists, such traces might be called 'biospheric fossils.'

This short excursion into the red planet's history may help explain why the term 'globosphere' would be useful. Before coining it, such a general term did not yet exist, while after having coined it, it may appear obvious. Furthermore, by examining globospheres one immediately realizes that a biosphere can be considered a special case of the more general category of globospheres.

The major questions for our biosphere's history then become: when, and how, did Earth's lifeless globosphere first turn into a biosphere, and what may have been its dynamics from its very beginning until our days?

Before seeking to answer those questions, let us first consider a few salient characteristics of globospheres. What do all the known planets and larger moons of our Solar System have in common? They are all ball-shaped, pulled together by the gravity that their mass exerts, which pulls this matter together and, in doing so, shapes it into globes.

Furthermore, the surfaces of such globes tend to be rather flat. And the larger those globes are, the stronger their gravity is that pulls them together, including the tendency to flatten their surfaces. From our human perspective, some terrestrial mountains may appear very tall. Yet seen from space, our planet looks like an almost perfect billiard ball. All of this is, of course, well known among astronomers and geologists.

The leveling of the surfaces of planets and moons is further stimulated by erosion processes, which are caused by the effects of free energy flows. Our Moon, for instance, rotates around its axis about once every month. This results in large differences in surface temperatures on its surface during the lunar days and nights, which expands and contracts accordingly. Over the course of time, this process has pulverized lunar rocks and turned them into moon dust, known as regolith. This is a very sticky material that tenaciously clung to the space suits of the astronauts who walked on the Moon.

On our nearest celestial neighbor, such erosion processes take place without the presence of water or wind. But when wind is present but no liquid water, such as on Mars' surface, this causes additional erosion, powered by the incoming sunlight. As yet, no planets or moons other than Earth are known that have liquid water on their surface or in their atmosphere that may further enhance erosion, even though some moons, such as Jupiter's moon Europa, may have liquid water below a top layer of water ice, while Saturn's moon Titan sports liquid hydrocarbons on its surface, the effects of which are adding to its erosion over time.

Furthermore, the unceasing bombardment of meteorites, large and small, also leads to further erosion, especially on the surface of globospheres that do not have an atmosphere, or only a thin one. They are, in consequence, unable to break up most of those incoming space projectiles. All those leveling tendencies tend to decrease the surface complexity of globospheres.

Flows of free energy and matter that emanate from deep within such globes also cause erosion. Yet they may also stimulate greater surface complexity, such as mountains, rivers, and deep canyons. On Earth, such effects include plate tectonics, volcanism, and earthquakes, as well as the associated movements of those tectonic plates.

Gravitational effects from outside may have similar effects, such as the tides on Earth caused by the gravitational attraction of the Moon and the Sun. This also includes Jupiter's moon Io's volcanism caused by the giant planet's strong gravity, which is kneading its nearby moon, and, in doing so, is producing those geological outbursts. Yet while sunlight exerts a leveling tendency on all globospheres because it stimulates erosion, within a biosphere it may also have an opposite effect, because it provides life with free energy to build structures, which may also hold soil together, and thus slow down erosion.

In addition, larger rocky planets are likely to have magnetic fields. That would result from their molten and spinning cores, consisting of mostly iron and nickel. It takes a large amount of time for a planet's interior to cool down. And the bigger such cosmic bodies are, the slower they cool down, and, in consequence, the more likely it is that they will have a magnetic field. This may shield the surface of larger planets from solar and cosmic radiation that is damaging to life, while it also protects their atmosphere from being stripped away by solar radiation.

Surely, there are many more characteristics of globospheres that could further be explored. But that is not the theme of this book. Here, we now seek to answer the following questions: when, and how, did the emergence of life turn Earth's early lifeless globosphere into a biosphere, and what may have been its first characteristics?

IN THE BEGINNING

In the beginning, about 4.56 billion years ago, the surface of the Earth would have been hot, desolate, and almost shapeless, and thus rather simple. It was not a place where you and I would like to have dwelled, or where our pepper plants would have flourished. In fact, as we will see much later in our account, it was only from as recently as about forty to twenty million years ago – a mere blinking of the eye in the biosphere's history – that our biosphere became sufficiently suitable for supporting our modern human lives.

In those very early times, there was no life yet, so, there was no biosphere either. While orbiting the infant and weakly shining Sun, our planet had emerged as a result of countless cosmic particles, large and small, colliding with each other and clinging together in the process. Over millions of years, this concentration process of cosmic matter produced our globe. The importance of concentration processes for our planet's history was emphasized by the US geologist Walter Alvarez in his book *A Most Improbable Journey: A Big History of Our Planet and Ourselves* (2016).

Where did all those cosmic projectiles come from? They were all leftover pieces from the Solar System's emergence, which was also a process of matter concentration. About 99% of that cosmic matter ended up in the Sun, while the rest either kept orbiting it or crashed into the emerging planets and their moons. Geologists call such concentration processes 'accretion processes.'

All those cosmic crashes onto the emerging Earth would have made our planet hot and molten. As part of that process, a glancing impact by a larger celestial body about the size of Mars would have sliced a considerable amount of matter out of our infant

Not exactly the beginning of the biosphere's history, but the Hoggar mountains in the middle of the Sahara Desert may look a little like such early landscapes. (Photo by the author, January 1981)

planet, part of which would have begun orbiting it, where it coalesced into our Moon. This collision would also have tilted Earth's axis with respect to the plane of its orbit around the Sun, in doing so producing seasonal changes on our planet's surface. Furthermore, the presence of a large Moon would have had a stabilizing effect on our planet's rotation around its axis, which enhanced the conditions for life to thrive.

What began happening inside the early Earth? The accumulated heavy and unstable radioactive chemical elements in its interior kept falling apart at their own particular velocities, as they always do. This process began to release large amounts of heat, which, together with the accretion heat, began to shape our planet through the turbulence that it created inside its body.

As part of that process, Earth's interior began to separate into several layers. Heavier chemical elements, mostly iron and nickel but also a few others such as gold, silver, uranium, and thorium, sank down and formed the core. Lighter chemical elements, mostly silicates, some of which were also radioactive, began to form the mantle covering the core, while the lightest rocky materials floated toward its surface. Those separation processes released energy, which made the already hot early Earth even hotter. All of that signaled the beginning of Earth's dynamics, driven by two principal sources of free energy from two different directions: the accretion, separation, and geothermal free energy from inside, and solar free energy from outside.

Did early Earth already have an atmosphere? The answer is probably 'yes.' Yet at a certain point in time, the emerging Sun is thought to have produced a sudden outburst of cosmic wind that would have blown most, if not all, of Earth's early atmosphere away into space. But after our central star had become more stable and stopped doing such violent things, a new terrestrial atmosphere began to take shape. This secondary atmosphere would have consisted mostly of carbon dioxide, nitrogen, water, perhaps some hydrogen, as well as a few other gases, but not yet any free oxygen, which currently makes up about 21% of the atmosphere.

Where did all those gases come from? The answers vary according to the academic disciplines. Geologists think that our planet's interior would still have contained such gases, and that over the course of time volcanism brought much of that to the surface, as it keeps doing, although today far less than in those earlier times. Yet according to astrobiologists (who still have to find any traces of extraterrestrial life), those gases kept arriving from space, most notably through comets.

In scientific descriptions of early Earth, the word 'climate' is rarely used, if at all. A similar situation can be found in descriptions of other celestial bodies within our Solar System. Today, not many scholars would speak of the climate on Jupiter, Saturn, or even Venus, if at all. Why not, one may wonder? The answer appears to be that none of those celestial bodies is supposed to contain any life. Yet on Mars, which still is a potential candidate for the existence of life, today or in earlier times, the word 'climate' is more often used, in doing so perhaps expressing the hope that on the red planet, life-forms, or ancient traces of them, may be found. Apparently, the word 'climate' implies the presence of a biosphere, or at least a potential one.

What would have happened on our planet's surface? In addition to the volcanism caused by Earth's hot interior, over hundreds of millions of years our planet kept being pummeled by celestial impacts, jointly called the heavy bombardment of meteorites. However, as of 3.8 billion years ago that barrage would have petered off. But it has not yet entirely stopped in our times, because every day many small meteorites keep entering Earth's atmosphere, while a few larger ones may hit us with larger intervals. Every year all of this cosmic stuff combined still adds, on average, about 40,000 tons of extraterrestrial matter to our planet. This is the equivalent of one hundred fully-laden Airbus A380 jumbo jets, the largest passenger planes currently in existence.

So, in sum: in the beginning Earth was a relatively uncomplicated hot rocky planet with a lifeless globosphere, perhaps much like Venus' globosphere today. As a result, geological and astrophysical knowledge would have been sufficient to describe and analyze our planet. And even though in those early times our planet's globosphere would have been complicated in its details – which is the case with almost everything that we can observe – its general structure would have been rather simple, because it was only conditioned by the relatively independent processes of cosmology and geology.

THE FIRST STEPS OF LIFE

All of that began to change as soon as the first life appeared on our planet. How did it get here? We do not know. It may have arrived from outer space. Yet according to the currently-dominant academic view, life probably emerged spontaneously on Earth's surface. This would have been preceded by a period of chemical evolution leading to the emergence of more complex molecules under the influence of geothermal energy, and perhaps also sunlight and lightning. In addition, comets and meteorites may have transported some, perhaps many, such molecular building blocks of life to our planet. From this primordial soup, the first life would have emerged, perhaps as early as four billion years ago. How that would have happened is still unknown, because traces of it have not yet been found.

What would the circumstances have looked like within which life first appeared? Around four billion years ago, the first salty oceans would have emerged, while there may also have been a few terrestrial areas sticking out of the water. Yet most of that land was probably not a good place for early life to emerge, not least because the ozone layer in the higher atmosphere did not yet exist, which today blocks most sunlight harmful to life. In the water, this liquid substance protects life against such damage. The ozone layer could only emerge after free oxygen had accumulated in the air. And that would have begun happening from two billion years ago as a result of the joint actions of the great many little green predecessors of our pepper plants.

Around four billion years ago, the secondary atmosphere was developing, containing water and other gases. And while around that time the cosmic bombardment was petering off, there would still have been considerable geological activity. Furthermore, the Sun was still shining relatively faintly, probably about 25% less intense than today. In addition, Earth had already a magnetic field. The earliest evidence of such a 'surprisingly strong' magnetic field is 4.2 billion years old, and comes from the oldest known rocks (Tarduno et al. 2020). So, apparently, such a magnetic field was already in place by the time life would have emerged, and would have shielded it against some cosmic harm.

Where on Earth would life have emerged for the first time? Perhaps in warm little ponds on land, to paraphrase Darwin's famous words, where reasonably good circumstances for the emergence of early life might have existed, especially if these were kept warm by geothermal energy? Yet although life's emergence in such circumstances cannot be excluded, today this is not considered a very likely scenario.

If life did not emerge in such places, where else would it have originated? This would have happened in the oceans near underwater volcanoes, where sufficient amounts of geothermal free energy and matter were bubbling up, but not too much. Those volcanic 'black smokers' would have been spewing out a great deal of that stuff, as some of them still do today. And some of that may have been used by early life to construct its first bodies and keep them going. As a result, in such situations a novel concentration process of matter and energy would have begun to occur, this time caused by emerging life.

However important it may be to reach a better understanding of how life emerged on our planet, this theme should not occupy us too much in our account, because we are more interested in examining what happened after life emerged on Earth, most notably how its actions began to transform our planet's earlier lifeless globosphere into a biosphere.[1]

What did early life look like? It must have been very simple, as all beginnings are. Those little pioneers must, therefore, have consisted of very basic single cells, each of them containing a minimum of biochemistry needed to survive and reproduce. Yet from those earliest primitive ancestors, all of life must have been descended, including you, me, and our pepper plants.

Unfortunately, we do not have any fossil remains of life dating back to that very early period. But we do have a rapidly-increasing knowledge of the hereditary materials of a wide range of living beings. All of that can be ordered into one single family tree, which appears to converge onto one single group of cells that would have lived some 3.5 billion years ago. But in all likelihood those cells were preceded by even more simple organisms. That is why I tend to be generous and state in this account that life may have evolved as early as four billion years ago.

How would those early ancestors have made a living? Those tiny cells would have emerged in areas in which there were sufficiently-large geothermal free energy and matter flows that could sustain them, but not so large that they would have destroyed them. Presumably, they would not yet have captured such free energy and matter actively. Doing so requires a specialized and more complex biochemistry, which might not yet have existed immediately after life's first emergence. As a result, such activities would not yet have been part of their primitive survival strategies.

Here, we witness a very general simple biospheric principle, namely that by necessity life is concentrated in areas where there are sufficient free energy and matter flows to sustain them. For the sake of brevity, this argument will not be repeated. But this principle has played a major role during the entire biosphere's history. It was only rather recently that a few species, including humans, began to concentrate free energy and matter supplies inside their dwellings.

Returning to the beginning of life: however uncertain the effects of early life within the emerging biosphere may have been, right from the start important

[1] My summary of the currently most likely scenario of life's emergence can be found in Chapter 4 of *Big History and the Future of Humanity, Second Edition* (2015).

biospheric patterns must have begun to emerge that subsequently made possible the extraordinary biospheric complexity of the most recent 650 million years or so. In that respect, the biosphere's history very much resembles human history. Starting from seven million years ago, all the human characteristics began to emerge that would make possible the similarly spectacular biospheric transformation that our species has brought about over the past 12,000 years. But before dealing with those more complex recent times, we first need to examine the long and far less complex early phase of the biosphere's history.

NONLINEAR GROWTH OF LIFE

To be successful in the struggle for life, those little ancestral cells must have possessed the ability to make copies of themselves. If not, they would soon have disappeared from the planetary stage as a result of the Second Law of Thermodynamics: the tendency of all matter to move toward greater entropy. So, when, and how, did life begin to reproduce itself? This appears to be unknown.

As soon as the first primitive life-forms began reproducing themselves faster than their death rates, their numbers must have grown. And as soon as the circumstances were sufficiently good for doing so, some of those early little cells may have begun to proliferate in nonlinear ways, much like the cells of our pepper plants did in the spring of 2017. As explained in the previous chapter, this can happen because one cell can engender two cells, which can produce four cells, etc., until the circumstances are no longer favorable for doing so. This is what most, if not all, microorganisms are still doing today. Such nonlinear growth would have led to a fairly rapid increase of their numbers. Unfortunately, it is unknown when such processes would have started for the first time.

THE EMERGENCE OF THE BIOSPHERE

What would have been the first effects of life that began to change our planet's lifeless globosphere into a biosphere? Nothing appears to be known about that subject, which, until now, may have escaped the academic attention.

Yet by creating their own bodies and performing all the novel biochemistry inside those little cells, while also excreting waste products, in Vernadsky's terms those first primitive organisms must have begun to put matter in unusual places as well as to shape it into novel molecular combinations. As part of that, they must have stored chemical free energy inside their bodies in the form of bio-molecules, in doing so producing the first accumulation of free energy caused by life. As result, the biosphere began to move a little further out of thermodynamical equilibrium than it already was, at least in that respect.

Those early activities must have begun influencing Earth's surface chemistry, but, of course, still in minute ways. As with all beginnings, those first biospheric effects must have been very limited. In those times, the biosphere was still in its very early stage of development, and any extraterrestrial visitors may not have foreseen that it would evolve into the extraordinarily complex biosphere that is familiar to us today.

It appears unlikely that evidence of such early biospheric effects will be found, because during the eons that followed, any possible fossils traces of them would have been wiped out by plate tectonics. Furthermore, in all likelihood the vigorously bubbling volcanic vents where early life supposedly lived would have erased most, if not all, the remnants of those effects very quickly anyway. Yet over the course of time, as the amount of biomass grew and life diversified, inevitably those first biospheric effects must have grown.

Even though we know virtually nothing about the emergence of those trends – typically a characteristic of 'first beginnings' – many of them must have started to occur in those early times. Interestingly, not a single account of the history of life or the biosphere known to me even mentions this issue, let alone that it provides an effort to outline those possible first biospheric trends in a general way.

So, let us try to do that. Three fundamental processes of early life influencing the biosphere can be distinguished: 1. through its sheer presence; 2. through its biochemical increasing complexity; and 3. through all the associated other actions and effects. As a result of these three fundamental processes, the free energy and matter flows on Earth's surface must have begun to change, while the first entropy produced by life was added to it as well.

Let us first very briefly examine the influence of life's sheer presence. Simply by being around, early life must have had some effects on the planet, because where life is, nothing else can be. In the beginning, those effects must have been minimal, while they are considerable today. In fact, in our times it is hard, if not impossible, to find any place on our planet's surface where there is no life at all.

The second fundamental way in which life began influencing the biosphere would have been through its increasing biochemical complexity. All those novel biomolecules would mostly have consisted of the chemical elements carbon, oxygen, hydrogen, nitrogen, and phosphorus, as they do today. Why those chemical elements? All of them are relatively light, while they are able to form so-called covalent chemical bonds, in which electrons are shared more-or-less equally between the chemical elements that form part of those particular chemical bonds. That works well for building larger molecular structures.

The chemically even more reactive elements such as sodium, potassium, calcium, magnesium, and chlorine occur in solution within cells in the form of electrically-charged ions, while they are hardly, if ever, incorporated within life's bio-molecules inside cells. Yet they play important roles in life's physiology.

Because all those chemical elements are comparatively light, they were concentrated on our planet's surface through geological processes, without which the emergence of life as we know it would not have been possible. Previous to our planet's existence, all those chemical elements had been forged inside large stars that had subsequently exploded. Such processes began to occur in cosmic history soon after the first big stars emerged, which happened about eight billion years before our Solar System emerged. Under the influence of gravity, those chemical elements had been concentrated in our emerging home planet as part of Earth's formation.

What about the third fundamental process in which life began to influence the emerging biosphere, namely through all its associated actions and effects? That will be discussed in the next section.

THE EMERGENCE OF COMPLEX CONTINUOUSLY INTERACTING FEEDBACK LOOPS

The emerging biochemistry introduced changes as to where some chemical elements were located. This signaled the beginning of life turning those chemical elements' globospheric cycles into biospheric cycles. In other words, those were the first such cycles in our biosphere's history that were influenced by both inanimate and animate processes.

However, instead of the term 'cycles' of chemical elements, in this account the word 'loops' will be used, because the first term suggests almost exactly-recurring events, while 'loops' may not do so.[2] In other words, the term 'cycle' is comparatively static, while the word 'loop' hopefully indicates a more dynamic process of change over time.

For example, while water molecules keep moving from the seas into the air and back, they never return to the same sea, or to the same air, simply because the biosphere keeps changing over time. As mentioned in the previous chapter, already the ancient Greek philosopher Heraclitus (c.540–480 BCE) recognized this fundamental aspect of reality by stating that "one cannot step twice into the same river." This deep insight was later reformulated by Plato (c.428–c.348 BCE), another ancient Greek philosopher, into the better-known phrase: "Everything flows, nothing stays the same."

The term 'loop' therefore appears more suitable to describe all the biospheric processes that never repeat themselves exactly. I am not completely sure to what extent the word 'loop' conveys this important insight sufficiently well. But at least it appears better than the word 'cycle,' and I am ready to yield to anyone who can come up with a better term for expressing such processes.

Because all biospheric loops tend to interact with each other while producing feedback effects, those very early life-forms would have set in motion the first processes which in this account will be called 'complex continuously interacting feedback loops influenced by life.' To avoid repeating such a reasonably precise, yet cumbersome, term, those processes will be abbreviated as 'complex cif loops.'

Because complex cif loops do not only interact within themselves but also influence other such loops, their emergence, change over time, and growing numbers have made our biosphere increasingly complex. This is a major reason, perhaps the major reason, why the biosphere has become so extraordinarily complex over the course of time.

To be sure, the idea that a great many such cycles, or loops, exist within our biosphere is not new at all. And neither is the notion that those cycles, or loops, influence each other, in doing so jointly producing greater complexity. Yet surprisingly,

---

[2] In the online Merriam-Webster dictionary, the word 'cycle' is defined as: 1. an interval of time during which a sequence of a recurring succession of events or phenomena is completed a 4-year cycle of growth and development; 2a : a course or series of events or operations that recur regularly and usually lead back to the starting point. More meanings are mentioned, but these are sufficient for our purpose: https://www.merriam-webster.com/dictionary/cycle (accessed December 29, 2020).

a general term that summarizes and expresses this idea does not yet appear to exist. That is why I decided to introduce the term 'complex cif loops' in this account.

Lifeless globospheres, including early Earth, usually also contain complex cif loops. They are all powered by external and internal sources of free energy. In the final analysis, free energy drives all complex cif loops, including the ones mediated by life. Yet within lifeless globospheres, such loops lack life's influences: its ability to harvest free energy and matter and do things with them that lifeless physical and chemical processes cannot do, including adapting to the circumstances through the process of natural selection. Within our biosphere, all of that has led to a whole novel set of emergent effects that have drastically changed it over the course of time as life expanded and diversified.

The early biosphere would have been characterized by the emergence of increasing numbers of complex cif loops mediated by life. Yet in those times, the presence of life would still have been limited, and, as a result, so would have been its biospheric effects. It is therefore hard to see that any early complex cif loops would already have produced global effects. But perhaps they did, and if so, this account would need to be changed.

Another part of life's early biospheric influences must have been the excretion of waste products, which added to their particular complex cif loops. But those earliest organisms presumably did not yet possess any capacity for intentionally going to other places. As a result, such movements would only have kicked in after living organisms began to acquire that ability. It appears to be unknown when that emerged. Yet as soon as cells learned to move around purposefully, this would have led to a whole new range of emergent joint effects within the biosphere, among which, much later in time, the actions of Charles Darwin's earthworms plowing the soil.

To be sure, all effects of life are part and parcel of complex cif loops. Living organisms influence, and change them by putting atoms and molecules in places where they would otherwise not have been. For the sake of brevity, this argument will not further be repeated. The reader is instead requested to keep this in mind.

Would early life have begun to influence the geothermal flows of free energy and matter that they depended on to survive and thrive? If so, only in minute ways. Those energetic outflows would have fluctuated mostly, if not entirely, as a result of physical processes, in doing so conditioning the growth and decline of the species that fed on them.

A SHORT HISTORY OF THE EARLY BIOSPHERE

Let us now try to sketch a possible scenario of life's early biospheric effects. As long as there were only a few little cells, those effects may have been small and localized. But we do not know for sure. In those very early times, there may have been a great many suitable places for life, wherever geothermal free energy and matter oozed out of the early Earth. And those pioneering little organisms feasting on them might quickly have learned to multiply, in doing so possibly producing numerous progeny in nonlinear ways within a relatively short period of time as soon as the circumstances permitted it.

By drifting along the sea currents from one volcanic vent to the next, those early pioneers may have spread around the globe very quickly. Those migrating microorganisms may not yet have captured geothermal free energy actively. But their diffusion within the emerging biosphere would have signaled the beginning of the Fourth Law of Thermodynamics: the tendency of life toward jointly maximizing the capture of the available free energy.

Meanwhile, geological processes were also making the biosphere more diverse. Both the land and the sea areas began to change over time, while becoming more diverse as well. In the waters, this would have ranged from shallow coastlines to the deepest oceanic troughs, while volcanism, or the lack of it, added diversity as well. Also land surfaces began to exhibit a growing diversity over time, varying from shallow coastal areas to the highest mountain tops. Both in the seas and on land, all those differences in elevations were unstable. While also in those times erosion tended to level Earth's surface again, volcanism did the opposite.

How did this growing geographical diversity influence life? In perhaps the shortest possible formulation mentioned by the Spanish astrobiologist Jesús Martínez-Frías in November of 2020 in an online lecture: growing geodiversity tends to lead to increasing biodiversity. The process of natural selection takes care of that by selecting for the sufficiently-fit species in all those areas. And because those regions were becoming more diverse, the same would have happened to all the early species that made a living in such places. Ever since their first beginnings lost in deep time, the joint processes of changing geodiversity and accompanying biodiversity have continued to exist up until today.

Those processes would have led to an increasing, and increasingly-diverse, number of biospheric zones, which, over the course of time, would connect to each other while sometimes separating again. The relative isolation of the landmasses of South America, Australia, India, and Antarctica after the breakup of the super-continent Pangea some 300 million years ago may serve as a more recent example of such a process of interconnection followed by separation.

All of this might have left chemical signals within the few rocks that have survived the onslaught of time. But finding them would require the knowledge of what those signals may have looked like as well as how to look for them, while for ascertaining a possible global distribution, we would also need to obtain reasonably representative samples of globally-distributed rocks from those early times. Those are major challenges but perhaps not entirely impossible.

Over the course of time, the emerging complex cif loops mediated by life must have had consequences for all living organisms, which would have been beneficial, neutral, or potentially damaging. And while more and increasingly-diverse microorganisms emerged, presumably their effects within the biosphere also became more varied. In other words, life's influence on the emerging biosphere also heralded the beginning of James Lovelock's concept of 'Gaia.'

This would have included, perhaps most notably, a possible selection for organisms that did not cause too much environmental damage to their own survival chances. It is currently unknown when, where, and how, such processes would have begun. And it is far beyond my current power and knowledge to trace all of that in any further detail for the entire biosphere's history. Yet such processes must have been happening for a long time. If we want to understand our biosphere's history better, outlining all those processes in more detail constitutes a major challenge.

In those early times, intelligent extraterrestrial visitors would have found it hard to detect any biospheric signals by sampling the water and the atmosphere, because those signal would have been overwhelmed by the strong geothermal processes, the incoming solar radiation, and the cosmic collisions that were then occurring on our planet's surface. This makes one realize that it is much easier to detect such tiny signals on a planet where there are far fewer geological or cosmic disturbances.

Today, such a situation can be found on the planet Mars. It is geologically almost 'dead,' while it is farther removed from the Sun than Earth, which, in consequence, leads to fewer solar effects within Mars' globosphere. Yet in June of 2019, the NASA Mars rover Curiosity detected a tiny signal of methane gas bubbling up from below the surface, while satellites orbiting the red planet had also detected similar signals. This may point to life below Mars' surface.

Such detections are not similarly easy on Venus' surface, which is situated much closer to the Sun while it also exhibits considerable volcanism. As a result, it is much more feasible to pick up such weak signals on Mars, even though doing so remains a major technological achievement. Yet picking up a potential signal of life's activities within Venus' relatively simple atmosphere, which is far less influenced by geological processes than its surface, is doable, as was reported in 2020.

Why would the discovery of methane gas escaping from Mars's surface be a potential signal for the presence of life? Because within the red planet's globosphere this gas is not in chemical thermodynamical equilibrium –, it should break down very quickly within the prevailing circumstances – those discoveries are currently leading to speculations about what may have caused that situation, most notably the potential presence of underground microbial life on Mars today.

This is not as farfetched as it may seem. Today, within our biosphere microorganisms can be found many kilometers below our planet's surface, where they are subsisting on geothermal energy. They may have migrated there, at least partially, under pressure of the changing biospheric circumstances that will be discussed below.

Similar processes may also have taken place on Mars. Some four to three billions of years ago, the Martian surface may have offered habitable circumstances for microorganisms. Yet as soon as the red planet lost most of its atmosphere and surface water, it became very dry and relatively cold, while its surface began to be hammered by solar and cosmic radiation almost unimpeded by its remaining thin atmosphere. That situation would have killed off most, if not all, life that may previously have existed there on its surface. Yet below it, microbes may have continued to exist until today, eking out a meager existence.

In sum: while the emergence of life on Earth also led to the biosphere's emergence, including the Fourth Law of Thermodynamics, the resulting biospheric effects would still have been tiny, while they appear to be still unknown. In those times, the geological and cosmic regimes would have been dominant in shaping the emerging biosphere, even though life's regime must have begun influencing it as well. Yet during the same period, the first fundamental biospheric processes must have emerged that have remained important ever since, most notably life concentrating matter in more complex shapes in places where it would otherwise not have been, while beginning to influence the complex cif loops of chemical elements.

As part of that situation, the evolving biospheric circumstances may have begun marginalizing, or even eliminating, all those species that could not sufficiently survive and adapt to those novel circumstances. In the beginning, such effects would have been very small, if they existed at all. But as the biosphere's complexity increased, so did such effects. For the sake of simplicity, in this account this argument will not further be repeated. The reader is instead requested to keep it in mind.

THE EMERGENCE OF ACTIVELY CAPTURING FREE ENERGY AND MATTER

Presumably between 3.8 and 3.5 billion years ago, microorganisms invented a novel survival strategy: actively capturing geothermal free energy and matter. In his book *Energy: Engine of Evolution* (2005), the Dutch scientist Frank Niele called this the emergence of the 'thermophilic (heat-loving) regime.' In our account, those novel organisms will be called the first primary captors of free energy and matter. It is unknown when the transition from passively to actively capturing geothermal free energy began to take place. But if life indeed emerged around four billion years ago, the period between 3.8 and 3.5 billion years ago may be a good estimate.

What happened as a result to the earlier passive captors of geothermal free energy and matter? Today, no microorganisms known to science are exclusively relying on the passive acquisition of free energy and matter. Such microbes may have been missed so far. And if indeed found, that would be very interesting. Yet today, all known life-forms are actively capturing free energy and matter. Apparently, this was such a successful innovation that those organisms outcompeted all of them who did not, with the result that most, if not all, those earlier microbes disappeared from the biosphere.

How would this capacity have improved the survival strategies of those microorganisms? Instead of relying on suitable gradients of geothermal free energy and matter to penetrate their bodies, the novel invention allowed such organisms to capture free energy and matter themselves. As a result, they could also begin to thrive in areas where the geothermal free energy flows were weaker, and thus potentially less damaging to their existence. And because Earth's interior kept cooling down, the outflow of geothermal free energy decreased over the course of time. Also in that respect, this novel biological invention would have worked to those microbes' advantage.

Yet the price to pay was a greater biochemical complexity inside their cells, most notably those biochemical reactions that were needed for capturing free energy and matter. Powering those more complex biochemical mechanisms must have increased the need for free energy. Yet ever since those early times, the active capture of free energy and matter has been an indispensable portion of the survival strategies of all known living beings. Apparently, the improved survival chances by doing so have outweighed the added costs until today.

Which sources of free energy were captured by those early pioneers? Most likely, those were types of chemical free energy that could be extracted by oxidizing simple molecules such as hydrogen sulfide and ammonia that had been created by geothermal processes. Today, many microorganisms are still living off such resources, which shows that it has been a remarkably successful survival strategy. However, such chemical extraction processes do not provide nearly as much free energy as sunlight, which immediate explains why, during the biosphere's history, organisms relying on geothermal energy have remained small. It was only by capturing sunlight that larger and more complex life-forms could emerge.

How was the capture of free energy and matter interlinked in those early times? In the process of capturing geothermal energy, there is an immediate connection, because microorganisms could only capture such free energy by harvesting energized matter. Only after capturing solar free energy had become an option – sunlight is pure energy – the processes of capturing free energy and matter became decoupled to some extent. But also the successful harvesting of solar energy could only be achieved while capturing carbon dioxide and water at the same time, because the captured solar energy needs to be converted into chemical energy stored in complex bio-molecules made by the cells from those simple inorganic molecules.

What may have been the effects of actively capturing free energy and matter within the biosphere? It would have led to an intensification of the trends mentioned earlier: a greater accumulation of free energy and matter inside their cells within more diverse bio-molecules. This implied increasingly concentrating atoms and molecules in places where they would otherwise not have been, while similarly influencing the biospheric complex cif loops related to those chemical elements. Yet in those times, life's effects within the emerging biosphere would still have been fairly limited. But as a result of the slightly increasing biomass and its growing effects, the biosphere began to move a little further out of thermodynamical equilibrium, at least in this respect.

DIVERSIFYING WAYS OF CAPTURING GEOTHERMAL FREE ENERGY

How did early life's survival strategies begin to diversify over time? While all the trends just mentioned began to occur, some primary captors of geothermal free energy may have started specializing in harvesting different types of free energy. For instance, some of them may have done so by oxidizing atoms, or molecules, that contained iron or sulfur, while others may have acquired different biochemical tricks to achieve similar results. Today, there is a considerable diversity of such microorganisms, which may have emerged as early as 3.5 billion years ago.

Whatever the case may have been, each and every innovation, when successful, leads to what is known in biology as an 'adaptive radiation:' the emergence of a range of similar, yet slightly different, species equipped with variations of that innovation. During such an adaptive radiation, usually only a limited number of successful species survives for longer periods of time. As 'our' Dutch paleontologist John de Vos observed, the process of innovations followed by adaptive radiations is a very general process that can be observed not only in the history of life but also in human history.

Curator John de Vos in Teylers Museum, Haarlem, the Netherlands, November 16, 2006, while giving us a private tour. (Photo by the author)

For a few such examples in recent human history, one can think of the emergence of smartphones, and, a little earlier, of personal computers. Those inventions quickly led to the emergence of a great many brands featuring slightly different products, of which only a few successful competitors have survived for longer periods of time, while they keep evolving. Likewise, among the early microorganisms that exploited geothermal free energy a range of slightly different survival strategies may have emerged. This would also have depended on the diversifying ecological biospheric circumstances within which such early organisms found themselves.

Today, all microorganisms that are living off geothermal free energy are jointly known as the *Archaea*, because all of them appear to have been descended from those earliest life-forms. Over the course of time, such microorganisms would have migrated to places where their favored free energy could be harvested, today ranging from hot geysers to layers deep below Earth's surface, while adapting themselves to those circumstances.

Such processes must have stepped up life's influence within the emerging biosphere, first of all by affecting the complex cif loops of chemical elements such as iron and sulfur that life began to use for obtaining free energy, while also intensifying their influence on the complex cif loops of other chemical elements such as hydrogen, carbon, oxygen, nitrogen, and phosphorus needed for constructing their cells. While also as a result of geological change all those complex cif loops kept changing, life's increasing influence must have led to more such complex biospheric processes with increasingly-intertwined effects.

THE EMERGENCE OF SECONDARY CAPTORS OF FREE ENERGY

A fundamentally-novel survival strategy emerged when organisms began to harvest their free energy and matter from primary captors instead of from inanimate nature. Those secondary captors of free energy let the primary captors do the hard work of concentrating inanimate free energy into bio-molecules, while they subsequently harvested it for their own purposes.

When, where, and how, would this have emerged? Again, we do not know. In fact, because of a lack of fossil evidence we may know very little about the history of those secondary captors as long as they remained small and single celled, or even smaller than that, such as viruses. Yet preying on other organisms must have begun a certain point in time. It may have started as early as three billion years ago, and perhaps even earlier than that. If so, this survival strategy may be very ancient.

What would have been the necessary conditions for the secondary captors' emergence? Obviously, they could only begin to make such a living after sufficient concentrations of first captors, or their leftover remains, had come into being. Before that time, there would not have been sufficient free energy and matter accumulated by life on which they could prey.

The emergence of the secondary captors may have been facilitated after primary captors began to exhibit nonlinear growth, because that provided the predators with more food. Such situations may also have influenced the growth of secondary captors in nonlinear ways. As usual, this novel survival strategy would have been followed by an adaptive radiation, leading to a greater diversity of secondary captors of free energy.

What would the survival strategies of early secondary captors have looked like? Their simplest form, known as scavenging, would have consisted of feeding on leftover bio-molecules of deceased organisms. Doing so may have required only a few biochemical innovations, because much of those microbes' internal biochemistry would already have been geared to metabolizing such molecules. But this could only work as soon as enough leftover bio-molecules had accumulated.

A second predatory survival strategy consisted of attacking living cells and feeding on them. This survival strategy worked perhaps the best as long as those secondary captors remained small and simple, while they could survive for longer periods of time without expending much energy. The easiest way of doing so consisted of shedding most of their own cell's content while repackaging a portion of its genetic material inside protein jackets, in other words: by turning into viruses. This allowed them to survive in the wild for long enough until they came across a living cell that could be attacked.

Once inside such a cell, the invader's genetic instructions would take over the cell's biochemistry and direct it to make as many copies of itself as possible. In doing so, such secondary captors could engage in nonlinear growth within such cells as long as the circumstances permitted it. After its resources had become exhausted, the doomed cell would fall apart and the newly-produced virus particles were set free to repeat this scenario.

This is what viruses do today. It is a cheap survival strategy, because it requires comparatively limited bodily complexity. It is therefore not surprising that there are staggering numbers and varieties of viruses today, which are infecting virtually all other life-forms. As so often, it is unknown when that started happening as well as how all of that may have evolved. But it may be a very ancient survival strategy. And because it appears impossible to group viruses within one single family tree, more likely than not viruses evolved independently several times. One would expect that viruses preying on primary captors evolved first, not only because those organisms have been around longer than secondary captors but also because they have a larger joint accumulated biomass.

As a result of the Second Law of Thermodynamics, over the course of time also viruses tend to decay. As always, this is speeded up by less favorable circumstances such as excessive heat, damaging chemicals, or electromagnetic radiation. For their continued survival, it is therefore imperative that viruses can either multiply in great numbers or that they can hitch a ride within the genetic material of their host cells. The latter is, for instance, what HIV viruses do. Those two options represent their major survival strategies.

And what about predatory cells eating other cells? One of their current ways of doing so is by wrapping themselves around other microbes and subsequently breaking them up, which is followed by harvesting their free energy and matter. It is assumed that from about two billion years ago, the larger so-called eukaryotic organisms emerged in such ways, but then as a result of only a partial breakup of the cells that were swallowed. If that latter view is correct, the survival strategy of 'eating' cells must be older than that. But it appears to be unknown when such predatory survival strategies would have emerged for the first time.

Much like humans have evolved defensive mechanisms against being robbed, so have all other life-forms. Doing so may have begun soon after the first secondary captors emerged that preyed on living cells. In his book *Plagues and Peoples* (1976), William McNeill called his dual process the problems of finding food while avoiding becoming food. Ever since the emergence of such predators, this has been an essential component of life's survival strategies. It was the beginning of the first arms' race among living beings, the end of which is not yet in sight. Also our pepper plants found themselves in such a situation. This makes one wonder what types of defensive strategies they may have evolved over the course of time, possibly including poisons for preventing their leaves to be eaten.

Even with such defensive measures in place, one may wonder whether predatory organisms sometimes wiped out the species that they were preying on. We do not know. While partial populations may go extinct as a result of such predatory actions, no entire species are known to have died out as a result of them. To the contrary, their populations usually appear to have bounced back. And if indeed secondary captors succeeded in terminating an entire species, for lack of food they might have gone extinct, too, unless they found another species to prey on.

Today, viruses occupy a special place within the food-capturing web. They may be the only organisms that are hardly ever pursued, if at all, by other species as a main source of free energy and matter, even though they may be ingested by accident. Viruses are simply too small to make their harvesting worthwhile, while they rarely, if ever, occur in sufficiently-concentrated numbers either. In other words, viruses find food but do not have to avoid becoming food. There are simply no organisms that have viruses for breakfast, lunch, and dinner every day. This makes one wonder whether becoming small and simple has been the viruses' preferred survival strategy of avoiding to become food, while at the same time offering a sufficiently-good way of finding it. All of that would have been honed, as always, under pressure of the process of natural selection.

Because attack and defense were added to life's survival strategies, some of those species became more varied and complex. In consequence, their bodies inevitably became more expensive to maintain in terms of free energy and matter, while they also began to produce more entropy. All of that must have influenced the early biosphere. But before discussing such possible effects, we first need to further examine the dynamism of life's interdependencies that resulted from the emergence of the secondary captors.

THE EMERGENCE OF THE FOOD-CAPTURING PYRAMID

The first appearance on the planetary stage of the secondary captors also represented the emergence of the food-capturing pyramid, which began to consist of dynamic and ever-shifting balances of power and dependency among its participants. As part of that, all the species involved would have been equipped with the inbuilt capacity for nonlinear expansion and decay, depending on the circumstances.

In this situation, the process of natural selection would have begun producing a widening range of microorganisms. Furthermore, the increasing competition for free energy and matter would have begun to put a premium on becoming more efficient. This would have driven all the microorganisms involved toward a growing diversity of increasingly-efficient survival strategies. This process has continued to exist ever since.

The rise of the food-capturing pyramid signaled, therefore, the beginning of life jointly making a more efficient use of the available free energy. This was a major novel trend in the biosphere's history. Instead of decaying after death, and, in doing so, returning to the universe in the form of infrared radiation whatever free energy was left in their bio-molecules, predatory species could now capture some of that free energy, dead or alive, before it disappeared into the cosmos and use it for constructing their own bodily complexity.

In other words, over the course of time the earlier food-capturing web changed into an increasingly-complex food-capturing pyramid, within which the captured free energy was used increasingly efficiently. This situation favored the emergence of new species equipped with novel inventions to capture free energy. This would have stepped up the effects of the Fourth Law: the tendency toward jointly maximizing the capture of the available free energy.

Throughout the history of life, those joint effects have very much depended on the available 'biotechnology' that allowed life to do those things: most notably, the molecular mechanisms and resulting biological structures that are used for capturing free energy and matter. During life's early history, this biotechnology must have been fairly basic and rather limited. But that does not invalidate the Fourth Law, which does not state that the joint capture of free energy by life is ever totally maximized, but instead that there is a tendency to jointly move toward that goal.

As mentioned in Chapter 3, by harvesting free energy from the primary captors, the secondary captors may have begun counteracting the Fourth Law to some extent. While doing so may have diminished the total amount of free energy that the primary captors were able to harvest, it did not affect the Fourth Law itself. It only meant that as a result of the emergence of secondary captors, the circumstances for jointly maximizing the capture of free energy may have become less favorable.

Yet if one examines what is happening in undomesticated nature today, including the little patches of plants growing in my Amsterdam neighborhood, one sees the Fourth Law in action almost regardless of how much free energy the secondary captors are tapping from the primary captors. For instance, when tree leaves are eaten by predators, other lower-situated green species may profit from that situation by capturing the sunlight that is no longer harvested by those unlucky trees. And this is only one example of a general process that must have emerged soon after the rise of the secondary captors.

The effects of the Fourth Law have been both facilitated and limited by the prevailing biospheric circumstances. Today, a lack of inorganic nutrients, for instance, insufficient water such as in deserts, or too much of it as in swamps, as well as temperatures that are either too low or too high, all present examples of how the capture of the available free energy can be limited by the prevailing circumstances. In principle, this must also have applied to those early times, even though the circumstances within which life then occurred were much more limited.

What about the numbers and diversity of all the microorganisms that have captured free energy during the biosphere's history? Even a beginning of an answer to that important question does not yet appear to exist. And how did the balances of power and interdependency within the emerging microbial food-capturing pyramid change over the course of time as a result of their evolving survival strategies? How many microbial survival strategies could be discerned over the course of life's history? And how did those survival strategies change during life's history, including those that went extinct? Again, with an awareness of great ignorance we are looking at a large field of academic inquiry that is of vital importance for understanding the biosphere's history.

And what about the differences in size between the first and the second captors of free energy? When secondary captors are smaller than their prey, such as viruses attacking microbes, or, more recently, viruses and microbes feasting on larger and more complex life-forms such as plants and animals, those little secondary captors tend to occur in large numbers, because many of them can live off the free energy and matter accumulated by a few larger individuals.

Yet as soon as the sizes of prey and predator become similar, such as microbes living off other microorganisms, or, much later in time, predatory animals such as lions attacking large herd animals like gazelles, antelopes, and zebras, by necessity there are fewer predators than prey, because if the numbers were equal, the prey would soon become exhausted. Such a difference in numbers becomes even more pronounced as soon as secondary captors become larger than the first captors.

Examples include cows eating grass, whales living off plankton, or dolphins, sharks, and birds eating fish.

By contemplating all of that, we are again jumping ahead in time. But from time to time we need to do so for demonstrating how trends that began small, with unknown origins, have evolved into more familiar trends. Without contemplating the beginning of a trend, even if it is entirely unknown, we may only partially understand its history.

WHAT MAY HAVE BEEN THE BIOSPHERIC EFFECTS OF THE EMERGING FOOD-CAPTURING PYRAMID?

The emerging complexity of the food-capturing pyramid and its external effects may have strengthened the first complex cif loops influenced by life. It may also have produced a widening range of biospheric areas, and, in consequence, of their biospheric effects. If all of that began to happen as early as between 3.5 and 3 billion years ago, life would still have been relatively rare. As a result, those biospheric effects would then still have been rather limited. But we do not know for sure.

How would the changing geological situation have affected the growth and spread of organisms? In the early phase of our biosphere's history, there was still a great deal of volcanism and other forms of geothermal activity. The occurrence of microorganisms would have been tied to such places. However, as a result of Earth cooling down over the course of time, the release of geothermal free energy began to decline. Yet this may have increased the number of milder volcanic environments that were suitable for early life, at least for a certain period of time. This situation may have facilitated its migration to such places.

As yet, no attempts appear to have been made to write even the beginning of a synthesis of the early biosphere's history in terms of those interrelated processes. At the current state of knowledge, it is therefore impossible to assess all of that with any degree of certainty. While, in all likelihood, in the early stage of the biosphere's history all those effects were still tiny, over the course of time, they must have been strengthening, diversifying, and becoming more globally distributed as a result of life's growth and increasing diversity.

THE EMERGENCE OF A NOVEL WAY OF CAPTURING FREE ENERGY: HARVESTING SUNLIGHT

Possibly as early as 3.45 billion years ago, a major new revolution in capturing free energy began to take place, as some microorganisms began to harvest solar energy through the process known as photosynthesis. In his book *Energy: Engine of Evolution*, Frank Niele called this the 'phototrophic (light-feeding) regime.' In our account, such life-forms will be called the first primary captors of solar free energy. If this process indeed began to happen 3.45 billion years ago, the biosphere would have been influenced by photosynthesis during 85% of its history.

How did that work, biochemically speaking? The solar free energy was captured by specific reaction mechanisms involving novel molecules and structures. As part of that, this free energy was added to the simple and abundantly-available molecules of water and carbon dioxide. In doing so, those inorganic molecules were combined into more complex and energized bio-molecules, first of all glucose, out of which all the other biochemical compounds were made. Those latter biochemical processes, jointly known as metabolism, are very similar today among all living organisms, and would, therefore, have evolved long ago, before photosynthesis existed. It therefore appears likely that the sunlight-capturing organisms added this novel survival strategy to their already-existing metabolism.

By beginning to capture sunlight, the biosphere's second major source of free energy, those pioneering microbes added an entirely new survival strategy to life's repertoire. Also in this case, the innovation would have been followed by an adaptive radiation, leading to a greater diversity of species that began to make a living off solar energy. In doing so, those novel species began to create a second, separate, food-capturing pyramid. Both pyramids, the older one based on geothermal free energy, and the newer one based on solar free energy, remained interconnected, because they existed within the same biosphere. In consequence, they were both influenced by the biospheric circumstances as they evolved.

Over the course of time, the food-capturing pyramid based on solar free energy would become dominant, because in large portions of the globe solar energy has been more abundantly available than geothermal free energy. In this account, for the sake of simplicity the distinction between those two food-capturing webs will not further be explored. Instead, they will jointly be called the 'food-capturing pyramid.' The reader is requested to keep this more complex situation in mind.

Why did this novel survival strategy emerge at that time? Was that perhaps related to changing biospheric circumstances? We do not know, but it might be the case. By that time, the geological and cosmic influences were becoming less intense, while solar radiation was increasing. This changing balance of free energy available within the biosphere may have stimulated the new inventors to 'follow the energy,' as Eric Chaisson called it in 2005 within the more general context of cosmic evolution. But whatever cause and effect may have been, the onset of harvesting sunlight has helped life not only to overcome a future shortage of free energy but also to harvest a source of free energy that was becoming more abundant.

The empirical evidence for this monumental change consists of fossilized bacterial mats dating back to at least three billion years ago that look remarkably similar to modern-day stromatolites. Today, such organisms consist of conglomerates of blue-green cyanobacteria that grow on top of rocks in the shallow waters along the western Australian coastline, on the interface between water and land. If those ancient fossils are indeed similar to their modern cousins, they must have been harvesting sunlight, too.

Because those early stromatolites were living together in the form of bacterial mats, this points to the earliest-known form of cooperation in the history of life. This also indicates that those organisms were a more advanced form of photosynthetic life than the original pioneers, which may have been floating around in the waters. Today, a great many of such single-celled photosynthetic organisms still exist. Yet because similar earlier single-celled organisms may not have left any traces in the fossil record, very little is known about them, if anything.

How did those tiny organisms evolve the novel survival strategy of capturing solar free energy? That is not yet completely understood. But in general terms, the ability

Not exactly stromatolites, but leading similar lives: green lichen growing on rocks in the harbor of our Amsterdam neighborhood, End of Sporenburg, October 2, 2016. (Photo by the author)

to capture solar energy must have evolved out of an already-existing biochemistry. As part of that, through the process of natural selection some microbes may have changed reaction mechanisms that were used to extract geothermal free energy into ones that could capture sunlight. That may sound farfetched, yet biochemically speaking this appears quite feasible (cf. Blankenship 2010).

How would this novel invention have spread? Today, the process of swapping genes among different microbial species is rather common. It accounts, for instance, for the rapid diffusion of resistance against antibiotics among microorganisms. As a result, the current microbial world can to a considerable extent be seen as one single common gene pool, which may have emerged soon after life originated. Because this common gene pool allows the swapping of biochemical mechanisms, this can quickly lead to the evolving, and spreading, of novel survival strategies.

How would those tiny life-forms have evolved the survival strategy of hanging together on rocks? This is similarly unknown. But biochemically speaking, it is even more feasible. For doing so, those organisms only needed to evolve a glue that could make their cells stick together. Such a capacity perhaps emerged earlier in time as part of a possible survival strategy of hanging together near volcanic vents.

The emergence of stromatolites represents the first example based on fossil evidence in which life began to cooperate, while also competing for resources.[3] The

[3] In my essay of 2018 'In Pursuit of Happiness: A First Exploration of Morality in Big History,' the processes of competition and cooperation are traced during the entire history of life, including that of humanity.

advantage for doing so would have been that those little cells could stick to favorable places, such as rocks in shallow waters, instead of being sloshed around to wherever the waters would take them. But the disadvantage would have consisted of sharing access to the needed limited resources, most notably the living space from where to harvest solar energy, which they tended to fill up. In consequence, stromatolites offer the oldest-available empirical evidence of the Fourth Law in action. Because the ocean waves would bring in the required inorganic molecules, their availability would not have been a major issue.

In our account, organisms hanging together will be called the first macro-captors of solar free energy, while single-celled organisms will from now on be called micro-captors. Yet compared to the much later, and far more complex, species that evolved a functional differentiation among their cells, leading to plants and animals, stromatolites have never done so. All their cells look alike, while they are all doing virtually the same, physiologically speaking. Over the course of time, hanging together has become a favorite survival strategy for a great many microorganisms, both in the oceans and on land, while living in groups has become a favored survival strategy for a great many complex life-forms including humans.

By forming what may have been the first microbial mats, those photosynthetic organisms became more concentrated. As a result, there may have been another price to pay, because those concentrated cyanobacteria would have become more susceptible to attacks by secondary captors such as viruses and perhaps also other microbes, if they already existed by that time. But even if such tiny predators were not yet around in those early times, they surely emerged later. Today, the Australian stromatolites of Shark Bay are under pressure of a great many viruses (cf. White et al. 2018).

To which other conditions did the photosynthetic organisms need to adapt? By relying on the globally-available solar energy, the photosynthetic microorganisms freed themselves from their bondage to places where geothermal energy could be harvested. But they did become dependent on a source of free energy that was intermittent, dictated by the rhythm of day and night. Because in those early times Earth was spinning much faster around its axis than today, there would have been about six hours of daylight and six hours of night time. Over the course of time, this day/night rhythm has steadily slowed down due to the tides, which, in their turn, are caused by the gravitational attraction by the Sun and the Moon. Those ever-changing water levels produce friction against the continents and, in doing so, slow down Earth's rotation around its axis.

Furthermore, there must also have been a changing influx of solar energy onto certain portions of our planet's surface due to the angle of Earth's axis with the ecliptic, the plane of its orbit around the Sun. Also in earlier times, this must have resulted in an annual summer/winter rhythm, while those changes very much depended on where one would find oneself on the globe. Because that angle appears to have changed very little over billions of years, the seasonal variations of the influx of solar energy would also have changed very little.

Those pioneering microorganisms must have evolved ways of overcoming the free energy shortage during the night, and the winter, while they may have preferred to live in the tropics where those annual changes are less pronounced. Biochemically speaking, storing sufficient energy for surviving the nights may not have been a great problem. It simply meant making enough energetically-charged bio-molecules to keep their biochemistry going all night long. Also our pepper plants were doing that.

As long as those early organisms sufficiently learned to do all of that, they would be fine. And because in those early times the nights were much shorter than today, this would have made it easier to bridge the resulting energy gap, in ways not unlike how the International Space Station is coping with that issue today. This huge structure circles Earth with some fifty minutes of light and forty minutes of darkness during every orbit. Powered by capturing sunlight with the aid of large solar panels, it stores a portion of that free energy in its large batteries to keep functioning on the dark side of Earth.

THE EMERGENCE OF BIOSPHERIC ECOSYSTEMS

The emergence of living communities near volcanic vents, and, later in time, the hanging-together survival strategy of stromatolites signaled the emergence of another major trend in the biosphere's history, namely of diversifying biospheric ecosystems. Over the course of time, those early assemblies of life-forms would have begun to consist of different species, including secondary captors. In doing so, they began to form the first ecosystems within which different organisms began to live together, all of them equipped with their own particular survival strategies. During the biosphere's history, a great many, often very diverse, ecosystems would emerge, decline, and disappear. These have all been self-organizing forms of ecological complexity, which can be seen as processes that change over time, even though the stromatolites, as we just saw, may have exhibited considerable continuity over long periods of time.

In the wet tropics, where sunshine tends to be the strongest, within those emerging ecosystems this has often led to great biodiversity, because in such areas there is so much solar free energy available. However, we do not know anything about this theme for early life. As a result, this subject will not further be explored here, also because tracing it throughout the biosphere's history would put us on a long Cape of Good Hope trajectory. As so often in this account, only the emergence of this important trend is mentioned, while no attempts will be made to trace its entire history.

Yet one aspect deserves some further attention. The life-forms that began to play an important role in ecosystems are known as 'keystone species.' Already in the year 1800, while crossing on horseback the dry and mostly barren plains of what is now Venezuela, the great Prussian naturalist Alexander von Humboldt observed that in small oases with little pools, *Mauritia* palm tree were growing. Those trees provided good circumstances not only for other plants and animals, among other things because of the shade that they provided, but they also offered a great many resources to human beings.

This raises the following questions: when, where, and how would keystone species have emerged in the biosphere's history? As far as we know, the early stromatolites

Alexander von Humboldt in his Berlin study, 1847. Farblithographie durch Bardtenschlager nach einer Vorlage von Eduard Hildebrandt. Farbendruck d. konigl lith. Instituts zu Berlin. v. Barth. (Original in possession of the author)

may have been the first such keystone species, to be followed by a great many others. Also this important theme will not further be explored in this account.

Which other novel effects may the early stromatolites have caused to their immediate environment? While they were hanging out on their rocks in the shallow coastal waters, they began to accumulate tiny sand grains that got stuck in their bacterial mats. This emergent effect would, much later in time, enormously increase their visibility as fossils, because sand grains are far less likely to decay over time than living cells. In fact, that is how we know about them. Their solitary cousins out and about in the oceans would not have caused such accumulations, and have, therefore, left far fewer recognizable traces in the fossil record, if any. Yet the solitary and solidary survival strategies both appear to have been quite successful, given that both types of species have continued to exist up until today.

THE FIRST MAJOR ACCELERATION OF BIOSPHERIC CHANGE

Ever since the primary captors of solar free energy emerged, they have jointly produced the first large biospheric changes occasioned by life. For over billions of years, with ups and downs, they slowly but surely began to change the presence of gases in the waters, the air, and on land. The most important changes were the steady removal of carbon dioxide, needed for making bio-molecules, while adding to their environment increasing amounts of free oxygen, which was a leftover product from photosynthesis. This began to influence the carbon and oxygen complex cif loops on a growing scale and at an increasing rate.

Furthermore, while this free oxygen was a waste product for the photosynthesizers, it did contain free energy. As a result, through the process of photosynthesis and the release of free oxygen, those solar energy capturing organisms began to power up the biosphere by adding free energy to it, in doing so moving it further out of thermodynamic equilibrium than it otherwise would have been.

In addition, the cyanobacteria may have been among the first, if not the first, life-forms to capture nitrogen from the air, where it has abundantly been available throughout the biosphere's history, and use it for making important bio-molecules, most notably proteins, DNA, and RNA. Capturing nitrogen from the air required considerable amounts of free energy, which had become sufficiently available thanks to the fact that those organisms were able to capture the abundant sunlight.

In doing so, the cyanobacteria also began to influence the nitrogen complex cif loop, including making this chemical element more readily available to other life-forms that could not perform this biochemical trick. Yet over the course of time, more microorganisms evolved able to do so. Much more recently, some of them, most notably perhaps the genus *Rhizobium*, would enter into a collaborative relationship with plants, supplying them with nitrogen extracted from the air in exchange for solar free energy and matter captured by the plants. But now we are again jumping ahead in time.

By changing the biosphere's chemical composition in such profound ways, life became a major player in shaping it. This has also produced large and important secondary effects which will be mentioned below, all of which have jointly turned the biosphere into a habitable environment for complex organisms such as you, me, and our pepper plants. In other words, the emergence of microorganisms more than three billion years ago able to extract carbon dioxide from the air and the water while dumping free oxygen into it has been of decisive importance for shaping today's biosphere.

Because the solar free energy that could be captured by life has been far more plentifully available than geothermal free energy, the effects of capturing it would also become much larger, while the similarly abundant availability in the waters of the major inorganic building blocks needed by photosynthetic life, most notably water and carbon dioxide, also favored the expansion of this process. While today about 70% of our planet's surface is covered by salty water, in earlier times that percentage would have been even higher. That situation allowed the early photosynthetic microorganisms to thrive in a great many places as soon as the circumstances permitted it.

The increasing extraction of the potent greenhouse gas carbon dioxide from the air while fixing it into biomass slowly but surely led to a gradual cooling of the biosphere. This would have been the first time in its history that life began influencing the biosphere's climate, which represented the onset of another major novel trend. This happened even though the Sun's energy output was steadily increasing, which would have produced a warming effect. Also the decreasing release of geothermal energy would have contributed to cooling the biosphere.

This cooling led to a greater diversity of temperature zones, which, in their turn, began to influence sea and air currents. Furthermore, this process would have changed the acidity of the oceans. While today's burning of fossil fuels by humans is putting more carbon dioxide in the air and the waters, and, in doing so, is producing an acidification of the oceans, the gradual extraction of this greenhouse gas by photosynthetic life must have had the opposite effect.

In sum: the expansion of photosynthetic life speeded up the effects of the Fourth Law, with corresponding growing effects within the biosphere, in doing so making it increasingly complex. And because the total biomass kept growing while it began to cover larger portions of Earth's surface, in those respects the biosphere became more complex, while moving a little further out of thermodynamical equilibrium than it would otherwise have been.

FURTHER PLANETARY FEEDBACK EFFECTS AS A RESULT OF PHOTOSYNTHESIS

The effects of photosynthesis began to produce even more feedback effects. As James Lovelock pointed out, the cooling of the biosphere caused by life's photosynthetic activities increased its capacity to keep liquid water on its surface instead of boiling it off into space. This important insight became a cornerstone of Lovelock's Gaia hypothesis, which states that life has kept our planet habitable over billions of years, in this case by helping to preserve sufficient liquid water within the biosphere. Over the course of time, incoming comets and volcanic outgassing would also have added water to it.

The ample supply of liquid salty water in the oceans inevitably led to evaporation, and thus to the distillation of water vapor that moved up into the air. Salts cannot easily evaporate in similar ways, and thus mostly stayed in the oceans. As long as those water clouds rained out over the oceans, not much happened. But as soon as they released their water over land, this made possible the emergence of fresh water pools and rivers.

It is unknown when such first fresh water pools emerged. But it is important to keep in mind that in contrast to the oceans, on land such fresh water reservoirs can exist in the form of pools, lakes, and rivers, which created novel biospheric circumstances. And because over the course of time Earth's landmasses increased, the fresh water reservoirs on land could also become larger and more numerous.

In those early times, there may not yet have been a great many microorganisms that could profit from fresh water on land, some hardy Archeans perhaps excepted. Yet by unceasingly slurping carbon dioxide from the oceans and the air, photosynthetic life has not only contributed to retaining liquid water within the biosphere, but it has also made possible the existence of abundant fresh water reservoirs on land, without which many forms of complex life, including humans, could not have emerged.

The unceasing photosynthetic activities also further influenced the chemical composition of the oceans. The free oxygen that was released as a waste product began to combine with soluble iron ions in the oceans, in doing so transforming them into insoluble iron salts that sedimented down to the sea floors. Over a long period of time, this produced the huge iron ore bands that are today mined by humans. After almost all the soluble iron in the oceans had been converted into such sediments, free oxygen began to accumulate in the oceans, while it subsequently also entered the atmosphere. This transition would have occurred about two billion years ago.

What would have happened to the interactions between the photosynthetic newcomers and the established captors of geothermal free energy? Because their sources of free energy were different, they may not have been living in the same places. In that respect, they would not have represented a direct threat to each other. Yet indirectly, there must have been interactions, because especially the presence of dissolved free oxygen in the water would have been poisonous to many older life-forms.

As a result, a process of natural selection would have kicked in that eliminated many of those ancient species, while others may have found refuge in places where free oxygen could not go, such as the deep oceans and inside Earth's rocky crust. And carried by the winds, some of them may also have moved onto land, in doing so perhaps beginning to reside in, or near, warm little fresh water ponds, especially during the times in which there was still little or no free oxygen in the air.

Such a scenario makes one wonder how the early photosynthesizers survived this environmental free oxygen crisis that they had caused themselves. Also they would have been descended from geothermal free energy captors. As a result, they would have shared a similar metabolism. For how many of them would free oxygen also have been a poison, and how many of them would have become extinct as a result, poisoned by their own waste products?

If the latter indeed happened, the process of natural selection would have allowed only those microbes to survive that could handle free oxygen sufficiently well, while the rest were either marginalized or totally disappeared. In Lovelock's terms, Gaia would positively have selected for those organisms that were sufficiently adapted to the new circumstances. Because so many of their descendants are still alive today, at least some photosynthetic newcomers must have been able to survive that onslaught. The idea of Gaia positively selecting for living organisms that were not causing their own destruction as a result of their biospheric effects is another major biospheric principle has occurred throughout its entire history. Again, the reader is requested to keep this in mind.

In sum: the innovation of capturing sunlight added a great deal of complexity to the biosphere. The microbes that fed on geothermal energy slowly but surely became less dominant, while the growing regime of life as a whole became more influential in shaping the biosphere, in increasingly complex ways. Many of those important trends would have begun relatively early, between 3.5 and 3 billion years ago, while they have continued to exist up to the present day.

INTERMEZZO: HOW LARGE IS THE BIOSPHERE?

How large is the biosphere? Most notably: how far does it extend down into Earth and out into space? As a result of photosynthesis, changes in the colors of Earth's watery surface would have begun to occur. Because it would take a long time before such joint effects would have become noticeable, let's say starting from about two billion years ago, that changed light would by now have traveled a distance of about two billion light years away from our home planet. In consequence, it could be argued that today, our biosphere's influence within the cosmos, faint as it may be, consists of a sphere with a radius of two billion light years. That would be a big sphere of influence!

This may sound farfetched, but that is exactly the way in which astronomers are looking today for habitable planets, namely by examining the light spectra of their presumed atmospheres. However, because the intensity of electromagnetic radiation decays with the square of the distance, in practice such observations are currently only feasible within much shorter distances, perhaps 200 light years or less.

However, this argument does not look entirely convincing. But it is not entirely unreasonable either. Much depends on how the biosphere, and the expanse of its influence, are defined. Would the biosphere solely consist of all the areas where life is present? Such a definition would limit it to a height of at most some 15 km above sea level, and perhaps as much as 10 km down into Earth. Or would we also include all the biosphere's emergent effects? In that case, we would need to include all the areas that have been influenced by life, including its emitted light.

Alternatively, the biosphere could be defined as the area influenced by life's complex cif loops. Such a definition appears to be more satisfactory, while it is very much in line with how James Lovelock defined his concept of Gaia on p.56 of his book *Gaia: The Practical Science of Planetary Medicine* first published in 1991. For a considerable portion of the biosphere's history, such a choice would yield a sphere between about 50 km above sea level, namely the upper level of the ozone layer, and a much more difficult to define limit below sea level, because we do not know how deep life's influences have reached below our feet. This definition would also further demonstrate the usefulness of the idea of complex cif loops influenced by life as a general concept, while it would yield an expanding biosphere over the course of time. It is therefore my preferred definition.

CHANGING COSMIC AND GEOLOGICAL INFLUENCES ON THE BIOSPHERE

Meanwhile, the cosmic and geological influences on the biosphere were also changing. The Sun's energy output had steadily been increasing, while the celestial impacts and the geothermal energy release had both been decreasing. This changed the face of our planet from a rather chaotic volcanic expanse into a more patterned surface, which began to be structured by the emerging process of plate tectonics. Powered by hot rising mantle material, our planet's outer skin began to consist of large plates that started to move around. Those nascent plates were sometimes colliding, while in other places they were breaking up and moving away from each other. In doing so, they began to produce the ever-changing geography of Earth's landmasses and oceans, which began to move around in patterns that are still being elucidated.

How long ago would the process of plate tectonics have emerged? The estimates range from three to two billion years before the present day. Before that time, Earth's surface and mantle would have been too hot and chaotic to allow that to happen. But the steadily-decreasing geothermal heat that was released from deep inside our planet helped to create the necessary conditions for plate tectonics to emerge.

As Walter Alvarez argued in his book *A Most Improbable Journey: A Big History of Our Planet and Ourselves* (2016), ever since the process of plate tectonics had started, it would have 'distilled,' concentrated, the comparatively lighter rocks such as granites into what became the landmasses, which as a result tended to grow over time. Because those granites weighed less than the heavier basalts, they tended to 'float' on Earth's surface, while the basalts ended up on the ocean floors.

All of that represented an important biospheric transformation, not least because it allowed materials from Earth's surface to be moved down into its mantle. This began to happen wherever landmasses and sea floors collided. As a result, the heavier basalts of the ocean floors, including any associated materials that had sedimented down onto them, began to move down into our planet's interior. And as soon as bio-materials, dead or alive, had also begun sedimenting down onto the ocean floors, those compounds were also transported deep into Earth.

Such influences would have been tiny in the beginning. Yet over the course of time, as a result of plate tectonics increasing amounts of sedimented bio-materials began to dive ever deeper into Earth. Crushed together and heated up under those harsh conditions, they lost most, if not all, their earlier characteristics. Yet after long periods of time, some of those materials would rise again to our planet's surface. This process led to what within geology is known as the 'rock cycle.' Over the course of time, such rock loops became an important aspect of the biosphere, where they would represent another major complex cif loop that was increasingly influenced by life.

Earth is currently the only planetary body in the Solar System known to have plate tectonics. Why would that be? Among others, the Dutch geophysiologist Peter Westbroek (1937–) gave the following answer. According to that view, life may actually have assisted in creating favorable conditions for plate tectonics to occur. The argument goes as follows. As James Lovelock had already argued, by sucking carbon dioxide out of the air, and, in consequence, lowering the biosphere's temperature, life has preserved copious amounts of liquid water on our planet's surface instead of losing it to the cosmos. Through the process of plate tectonics, some of that water would have been transported down into Earth's interior. This subducted water may have acted as a lubricant for the tectonic plates, in doing so facilitating their movements while assisting them to keep this process going. If correct, this would have represented the emergence of a major novel biospheric complex cif loop influenced by life.

This argument is strengthened by the absence of plate tectonics on the neighboring planet Venus. It has more or less the same size as Earth, and would, in consequence, have had a similar geological history. Yet because Venus is situated closer to the Sun, and as a result has a hotter surface, our sister planet has lost all its water, while it does not support any life. So, why would Venus not have plate tectonics? Would Venus' globosphere perhaps not provide sufficiently-good conditions for plate tectonics because it lacks life that would have assisted in preserving its surface water, and, in consequence, would have helped create plate tectonics within a Venusian biosphere?

Whatever the situation may have been, the cooling down of Earth's interior in combination with the emergence of plate tectonics made possible a greater geographic variation. This process did not only produce the differences between the landmasses and the oceans, but also the growing diversity both on land and in the waters, such as mountains and deserts, deep ocean troughs and shallow coastal waters. And as usual, life must have responded to this growing geodiversity by an increasing biodiversity, which was caused by biological innovations and the following adaptive radiations under pressure of the process of natural selection. Yet because most material remains from those early times may have been erased by plate tectonics, by erosion, as well as by life's later actions, very little is known about those early circumstances.

COMBINING SURVIVAL STRATEGIES

Between 2 and 1.5 billion years ago – so, starting at the last 50% of the biosphere's history – a new type of living cells emerged that was considerably bigger and more complex than their predecessors. Those larger cells contained a compact nucleus where the molecules carrying the cells' genetic inheritance became tightly packed. That is why those cells are called eukaryotes, because in Greek this means 'entities with a clear nucleus.' Those novel cells also began to contain other little structures called organelles, which began to specialize in executing certain tasks.

Because more complexity requires more free energy and matter to keep going, this raises two fundamental questions, namely: how could that have happened, and, why would all of this have been sufficiently advantageous for surviving the struggle for life? For the beginning of an answer, let us look at what may have been the two most important novel organelles. Those are the so-called chloroplasts, which are solar energy-capturing units, and the mitochondria, which are energy-processing units using free oxygen. Not all the eukaryotic cells possessed both types of organelles. But they would all have nuclei and mitochondria. This immediately suggests that cells containing nuclei and mitochondria would have emerged first, while some of them, but not all, later also acquired chloroplasts.

In this process, the obvious difference emerged between cells that contained chloroplasts and those that did not. This signaled the beginning of a divergence among living organisms that would, much later in time, lead to novel and more complex primary captors of solar free energy such as plants, and similarly more complex secondary captors of free energy such as animals. This early development and its later consequences immediately explain why our pepper plants sported branches holding many green leaves, the cells of which all contained a great many chloroplasts, while you and I have only a few limbs that are not carrying any solar arrays.

Those differences became more pronounced as soon as eukaryotic cells began to form multicellular structures, in doing so hanging together in ways that were similar to what the stromatolites had already been doing for one billion years or more. Based on genetic studies of today's eukaryotes, all of that would have begun to emerge around 1.5 billion years ago. Stated in more general terms, this represented the emergence of a novel range of more complex species that were, in consequence, equipped with more complex survival strategies.

Hardly any fossil evidence, if at all, appears to exist attesting to those changes. This may indicate that the initial biospheric impact of those novel organisms would not have been very large. However, plate tectonics and erosion may have erased such possible traces.

Between 1.5 and 1 billion years ago, among the eukaryotes also the innovation of sexual reproduction would have emerged. Yet as mentioned earlier, in all likelihood the simpler prokaryotic microorganisms would also have been able to exchange genetic materials, quite likely more easily than eukaryotes. The emerging eukaryotes would not have been able to exchange their genetic materials easily with other similar species anymore, because their nuclei, where their genetic material became tightly packed, were more complex and better protected. They became like little castles, and were therefore more insulated from each other.

The emergence of sexual reproduction among eukaryotes may, therefore, have been a reaction to that restricting situation, favoring those organisms that could exchange genes in sexual ways, because that provided a greater range of genetic variation which was helpful to survive the process of natural selection. By creating stricter barriers for the exchange of genetic materials among species while facilitating this within a certain range of them, the rise of eukaryotes would have signaled the emergence of the first species that could more clearly be defined as such by biologists.

How did the first eukaryotes emerge? It appears likely that this happened after certain predatory microbes encapsulated their prey. But instead of completely digesting it, they allowed some of them to continue to exist inside their cells in altered ways, while engaging in forms of biochemical cooperation. The idea that such processes would have happened immediately points to a previous existence of secondary captors able to do that, as well as to the existence in those times of specialized, yet more simple, microorganisms that had become experts in either capturing sunlight or processing free oxygen for powering metabolic processes.

So, apparently the microbial predecessors of chloroplasts would have been primary captors of free energy such as the cyanobacteria, while the mitochondria might have evolved out of secondary captors of free energy that had learned to use free oxygen as part of their metabolism. Doing so releases far more free energy stored in bio-molecules. In consequence, this innovation allowed a more efficient use of the available free energy, which made it possible to support more cellular complexity.

In other words, the eukaryotes would first have evolved by ingesting more simple secondary captors of free energy that turned into mitochondria. Subsequently, some of them would also have encapsulated little primary captors of solar free energy, which became the chloroplasts. Ingesting the predecessors of chloroplasts must have been tricky, because the free oxygen released by them inside such combined cells as a result of their photosynthetic activities may have had damaging effects. Such cellular fusions may, therefore, not have occurred successfully very often. Yet the mitochondria needed free oxygen for functioning well, and the nearby production of it by chloroplasts may have entailed an advantage. By combining those survival strategies within one single larger and more complex cell, the diversifying eukaryotes represented remarkable innovations in capturing and processing free energy and matter.

Why would this have happened during that particular period in time? The answer seems obvious, and is well-known. Those larger and more complex cells became possible only after enough free oxygen had accumulated in the water that could power those more complex cells. And that could only happen after most, if not all the inorganic iron in the oceans had been oxidized. So, in sum: the appearance of free oxygen in the oceans would first have led to simple bacteria, the predecessors of mitochondria, that were able to use it, and subsequently to the ingestion of some of them by the emerging eukaryotes.

As mentioned, using free oxygen to break down bio-molecules yields far more free energy. In consequence, the appearance of free oxygen in the water, and later also in the air, signaled a most important change in the biosphere's history, because for the first time this allowed the emergence of complex organisms that required much more free energy to exist and prosper. In his book *Energy: Engine of Evolution*, Frank Niele characterized this change as the 'Oxo-Energy Revolution.'

The importance of this development should not be found in terms of any great biospheric change that immediately happened during those times. Around 1.5 billion years ago, an intelligent observer from outer space would hardly have noticed such impacts, if at all. But while, as usual, the beginnings of this novel trend were small and seemingly insignificant, over the next 1.5 billion years eukaryotic cells would develop into a great many types of complex life including you, me, and our pepper plants. In other words, those pioneering eukaryotes became the building blocks of all more complex organisms.

Today's more simple microbes are called prokaryotes, 'before having a nucleus' in Greek. They cannot form similarly-complex living beings. The most complex structures they can make are microbial mats. Why not? An important part of the answer appears to be that eukaryotes can pack much more genetic material in their cells than prokaryotes. In consequence, they can contain much more genetic information that codes for their greater complexity. Yet greater complexity has a price tag attached to it, namely a larger need for free energy and matter. The newly-absorbed chloroplasts and mitochondria, with their far greater capacities for capturing free energy and consuming it, allowed the eukaryotes to do all of that.

How would the addition of eukaryotes to the food-capturing pyramid have affected the entire pyramid? No studies about this subject are known to me. One would expect that those larger eukaryotic cells soon became a food target for the earlier and smaller secondary captors, most notably perhaps viruses, and especially so when some of the eukaryotes began to turn into multicellular structures, because such further concentrations of free energy and matter would have been beneficial to microbial predators including viruses.

And how would the novel eukaryotes without chloroplasts, which as a result were secondary captors, have found ways to harvest enough free energy and matter to maintain their larger cells? Also this may still be unknown. Yet one may assume that this novel situation would have led to dynamic processes of attack and defense among all species involved, in doing so leading to a larger, more complex, and more dynamic food-capturing pyramid that kept evolving over time.

And what about the effects within the biosphere by all of this? Very little, if anything, may be known about this subject either. No new major novel complex cif loops appear to have been added by the early eukaryotic presence, with the exception of the loops of a few heavier chemical elements that began to be used as so-called spore elements. In very small quantities, those chemical elements might have begun to play a role within the more complex eukaryotic biochemistry. But their biospheric effects would have been tiny, and would remain so during most, if not all the biosphere's history. Yet one may suspect that all the effects combined resulting from the emergence of eukaryotes made the already-existing complex cif loops influenced by life a little more complex and dynamic.

LIFE KEEPS COOLING THE BIOSPHERE

How did life's increasing effects change the biosphere from 2.5 billion years ago until larger and more complex life emerged around 540 million years ago? This long period is known as the Proterozoic. The most important effects by life came as a result of photosynthetic organisms slurping carbon dioxide from the air and the water, while releasing free oxygen into it. After such organisms had died, their growing numbers left increasing quantities of bio-molecules within the biosphere, which, apparently, were not entirely consumed by secondary captors. Instead, some of those bio-materials accumulated over time into what today are called the 'Proterozoic oil deposits.' Such deposits, the oldest of which would date back to about 1.3 billion years ago, were left in many places, such as what are now Canada, the Middle East, Russia, and Australia.

This represented an intensification of life's efforts of moving atoms and molecules to places where they would otherwise not have been, as well as a similar intensification of biospheric processes that preserved of some of those bio-molecules for longer periods of time by burying them under layers of sediments. As we saw earlier, among the oldest-known deposits of life are the fossil remains of stromatolites, which, however, contain very few bio-molecules, if any, while tiny traces of even earlier life may have been preserved as well. Yet none of those very early deposits contains substantial amounts of free energy. By contrast, the Proterozoic oil deposits obviously do so, which is the principal reason why humans have been extracting them.

Also some earlier Archaean bacteria living off geothermal free energy had begun to release bio-molecules within the biosphere that contained free energy, most notably methane gas that for them was a waste product, much like free oxygen was so for their photosynthetic cousins. Over the eons, some of that methane gas has accumulated below Earth's terrestrial surface, and also in dissolved form in cold water near the ocean floors. As a result of all those accumulation processes, life jointly further powered up the biosphere, in doing so pushing it farther out of thermodynamic equilibrium than it otherwise would have been.

Life's steady removal of carbon dioxide from the air and the water kept cooling the biosphere. Over the past one billion years – the most recent 25% of the biosphere's history – this cooling produced several cold periods known as Snowball Earth events. During those times, most of the planet's surface would have been frozen. Apparently, as a result of life's joint actions the biosphere's climate began to fluctuate over longer periods of time, shifting between warm and cold periods. Especially those cold periods would have posed severely limiting conditions on the possibilities for the food-capturing pyramid to survive and thrive. As a result, it diminished in size but perhaps not in diversity. In fact, this was a situation in which life began to do almost the opposite of Lovelock's Gaia hypothesis, namely undermining the conditions for its own survival and prosperity.

How would the biosphere have escaped from such a frigid destiny? Much like the Snowball Earth theory itself, this issue is currently very much under discussion, and unequivocal answers do not yet appear to exist. But it looks as if the output of geothermal free energy from deep inside the Earth may have come to the rescue. How would that have worked?

During a Snowball Earth event, the frozen state of much of Earth's surface and the resulting strongly-reduced photosynthetic activities implied that only limited amounts of carbon dioxide were further extracted from the air and the water, while any addition of this gas to the biosphere by geological processes would not immediately be sucked up by life. It has been suggested that increased volcanism near the end of such cold spells would have added carbon dioxide to the biosphere. This would have created a stronger greenhouse effect, which led to a warming up of the biosphere. Yet it still needs to be explained why the subsequent increasing photosynthetic activities of green life would not have reversed this warming trend very quickly.[4]

Although much uncertainty still exists, the last Snowball Earth event would have ended around 650 million years ago. Why did not any more recent such events occur? Would the emergence of more complex life starting from around 540 million years ago and its effects within the biosphere perhaps be implicated? Would this have kept the biosphere at temperatures that were more favorable for life to flourish? If so, any biospheric mechanisms that might have regulated that situation do not appear clear to me.

Yet regardless of those tricky issues, after the last Snowball Earth event had ended, the process of photosynthesis kept sucking carbon dioxide out of the air and the water. In doing so, it reduced its concentration to perhaps the lowest-possible levels that allowed green life to keep going. And even while today humans are putting large amounts of carbon dioxide back into the air, leading to a warming up of the biosphere, the current total carbon dioxide concentration in the air is still only some 0.04% of all its gases, while free oxygen currently amounts to about 21%. Without the persistent activities of green life, today there would not have been any free oxygen at all within the biosphere, while the carbon dioxide concentration in the air would perhaps have been as high as 90%, if not higher.

In fact, by continuously lowering the carbon dioxide concentration in the air, the photosynthetic organisms created a problem for themselves, because they turned it into a scarce resource. In consequence, those life-forms had to learn to suck ever harder to obtain the needed carbon dioxide. Doing so inevitably implied using more free energy to achieve that goal. It is unknown to me how high those energy costs have been. But this issue may contribute to explain why green plants tend to produce such large green surfaces. While our pepper plants were doing that in 2017 and 2020, we did not observe any huffing and puffing caused by their carbon dioxide suction efforts, because plants do not produce any such noises. We only saw their rapidly-increasing foliage, while we could not see what the captured free energy was used for or what they were otherwise doing. That is why such processes may easily escape the attention.

Because of this situation, inevitably a competition among species must have emerged, favoring those species that could obtain their carbon dioxide sufficiently

---

[4] As so often in this account, given its limited size, it is impossible to do justice to all aspects involved. This also applies to Snowball Earth. An excellent overview of the academic discussion of about ten years ago can be found in the book *The Goldilocks Planet: The Four Billion Year Story of Earth's Climate* (2012) by the UK geologists Jan Zalasiewicz and Mark Williams (pp.21-52).

efficiently. This would also have included more efficient ways of capturing solar free energy. Any innovation that would improve such processes must have offered a survival advantage, and would, therefore, have led to adaptive radiations. Apparently, our pepper plants represented such a sufficiently-successful species. This development also represents an example of resource extraction by life leading to scarcity, and thus requiring adaptations for securing its continued existence. It may not have been the first such occurrence. Quite possibly, similar situations had already emerged earlier in time. But none of that appears to be known to science.

Whatever exactly may have happened between 2.5 and 540 million years ago, the general trend has been, with ups and downs, an intensification of life's influences within the biosphere.

GEOLOGY AND LIFE BEGIN TO INTERACT MORE INTENSIVELY

Also starting from around 2.5 billion years ago – which, or may not, have been a coincidence –, the process of plate tectonics began to produce the so-called 'super-continent cycle,' during which virtually all the landmasses joined at certain periods in time, only to become separated again many millions of years later. Although much uncertainty still exists, between 600 and 540 million years ago such a super-continent, called Pannotia, would have existed.

Whatever may have caused the super-continent cycle, those large-scale events posed certain conditions on the possibilities of life to flourish, even though in those times almost all living organisms were still to be found in the waters and not on land. For instance, one single landmass tends to have a shorter shoreline than many broken-up pieces of land. And the shallow seas surrounding landmasses usually provide good living conditions, not least because water running off the land brings in the needed inorganic minerals. As a result, a super-continent would have lowered the survival chances of the food-capturing pyramid because of its smaller shorelines and correspondingly more limited shallow seas.

Yet starting from perhaps as early as 1.2 billion years ago, and perhaps more clearly so after the last Snowball Earth event had ended about 650 million years ago, the landmasses may no longer have been completely lifeless, because in the meantime, the photosynthetic effects of life had begun to create a global sun shield that began to protect life on land against the onslaught of ultraviolet solar radiation.

This would have worked as follows. The continuing release of free oxygen into the air had led to the emergence of an ozone layer at an altitude between 15 and 35 km above Earth's surface. The gas ozone consists of three oxygen atoms bound together, instead of the far more common molecules of free oxygen that consist of only two oxygen atoms. This ozone gas forms out regular oxygen molecules higher up in the atmosphere under the influence of sunlight. It is chemically unstable, and tends to falls apart fairly quickly. For an ozone layer to exist, it does, therefore, need to be replenished by the continued influence of ultraviolet solar radiation. The ozone layer prevents a considerable portion of the Sun's ultraviolet radiation from reaching our planet's surface. This situation represented another novel complex cif loop added by life to the biosphere, which would not have existed without life's persistent photosynthetic activities.

In consequence, the growing ozone layer considerably improved the conditions for life on land, first of all because it no longer needed the protection against solar ultraviolet light that the aqueous surroundings provided. Yet it is unknown when life first moved on land, and how that may have happened. More likely than not, the first hardy colonizers were microorganisms carried there by the winds. Some of them may have been able to capture solar free energy and settle on land. And in areas where there was enough geothermal energy available, microbes able to capture that type of free energy may also have made such a jump. If so, all of that would have produced considerable biospheric changes, including further effects of the Fourth Law.

THE EMERGENCE OF THE FIRST COMPLEX MULTI-CELLULAR LIFE-FORMS

At the beginning of the last 16% of the biosphere's history, around 635 million years ago, the emergence of novel life-forms began to reshape the biosphere in fundamental ways. This happened soon after the last Snowball Earth event had ended while the super-continent of Pannotia was taking shape. In those times, the first larger and more complex organisms emerged that left traces in the fossil record. All of them would have consisted of eukaryotic cells that jointly formed soft bodies with specialized organs. The period of their existence, known as the Ediacaran, lasted between 635 and 542 million years ago.

All those species would have lived in the water, while they looked like precursors of plants. Yet it is not clear whether they were indeed early plants. The reason for that is that no one knows on what types of free energy and matter those organisms would have fed, inanimate or produced by life, in other words, whether they were primary or secondary captors. The currently-dominant interpretation is that those life-forms may have harvested organic nutrients from the oceans, which would have been plentiful at that time.

The bountiful presence of such organic molecules in the waters represented a major biospheric change. Yet it appears to be unknown for how long such accumulations resulting from decaying single-celled organisms had existed as well as how plentiful they may have been. If the Ediacaran organisms indeed fed on such organic molecules, they must have been scavengers, secondary captors of free energy and matter. In doing so, would those more complex newcomers perhaps have faced competition from earlier-established single-celled secondary captors? If so, what would have been the biospheric consequences? None of that appears to be known.

What would have been the effects of the Ediacaran species on the food-capturing pyramid as well as within the biosphere as a whole? This may also be unknown. One may suspect that the Ediacarans became a food target for little secondary captors such as microbes and viruses. If so, that would have led to more, and more complex, interdependencies within the food-capturing pyramid, while it would also have signaled another step of life in jointly making more efficient use of the available free energy.

Furthermore, some novel Ediacaran species may have begun to act as keystone species within ecosystems that were becoming more complex. Other Ediacaran species may have migrated to the land, most notably perhaps to warm little ponds, in doing so possibly adding to the diversity of life on *terra firma*. If so, they might have been accompanied by solar energy harvesting microorganisms as well as by microbes and viruses that fed on both of them, in doing so potentially producing the first food-capturing pyramids on land as well as the first terrestrial ecosystems. But again, none of that appears to be known, so this remains speculative.

Yet however important the rise of the Ediacaran species may have been, any novel biospheric effects caused by them were probably dwarfed by the expansion of the food-capturing pyramid as a whole after Snowball Earth had ended as well as by the emergence of super-continent Pannotia. All those changes must have produced enormous biospheric effects. By comparison, the Ediacaran influences may have been fairly limited. But we do not know for sure. As so often in our attempts to reconstruct the biosphere's history, we are dealing again with considerable ignorance.

THE EMERGENCE OF A MORE COMPLEX FOOD-CAPTURING PYRAMID

Around 540 million years ago, spectacular novel biological changes began to take place that are jointly known as the Cambrian explosion of life-forms. During a period of only a few million years – a fraction of 1% of the biosphere's history – the earliest-known complex animals and plants began to leave their traces in the fossil record. In terms of the biosphere's history, the period since the beginning of the Cambrian period until today is relatively short. It represents no more than the last 13.5% of the biosphere's history.

This was the first time in the biosphere's history that large organisms emerged. All of today's complex life is descended from those species, including you, me, and our pepper plants. It is far more difficult to assess what happened during that period to the microorganisms, because they left far fewer traces in the fossil record, if any. The novel Cambrian species were all complex multicellular eukaryotic life-forms. They showed a considerable functional differentiation among their cells. In other words, cells became specialized, while they remained part of one single individual life-form. Conglomerates of such cells began to create specialized organs ranging from eyes and legs to roots and leaves.

In other words, the Cambrian period heralded the rise of the first larger and more complex primary, secondary, and tertiary captors of free energy. As part of those changes, a clear difference began to emerge between the already-existing micro-captors and the newly-emerging macro-captors of free energy. To some extent, such differences had already appeared in the Ediacaran period. But during the Cambrian, they became far more pronounced. All those changes combined added a great deal of complexity to the food-capturing pyramid, while they offered another example of the Fourth Law in action. As a result, very quickly the biosphere became much more complex as well.

It is unknown whether those novel species were descended from the earlier Ediacaran life-forms. The current dominant view is that most, if not all of them were not related in such ways. Yet in 2020 it was reported that fossil traces of a tiny novel species named *Ikaria wariootia* were discovered dating back to 555 million years ago, which may have been our earliest-known complex ancestor, or relatively closely related to it (Evans et al. 2020.). If that view is correct, the Cambrian complex life-forms may indeed have been descended from Ediacaran species. Yet whoever the ancestors of the Cambrian plants and animals may have been, within a few million years a great variety of complex plants and animals emerged. Because their innovations led to adaptive radiations, this produced an increasing diversity of complex life-forms over the course of time.

While considering this period, usually most of the attention is focused on the spectacular and often weird-looking animals, at least to modern human eyes. The less spectacular Cambrian plants tend to receive far less attention, while the rest of the food-capturing pyramid, all the microorganisms, are usually almost completely ignored. Yet during the Cambrian period, all those little organisms must also have been present. In our account, this bias will be called Bias # 8. The fact that during the Cambrian period suddenly so many complex animals appeared that were all feeding, directly or indirectly, on the free energy and matter harvested by primary captors, indicates that there must have been a great many of them capturing solar, and perhaps also geothermal, free energy.

All the complex animals were, by definition, secondary or even tertiary captors of free energy. Tertiary captors are life-forms that feed on secondary captors. They include complex animals eating other complex animals as well as microbes and viruses feasting on animals. Today, also animals exist that feed on both primary and secondary captors, most notably perhaps our own species, since many of us eat both plants and animals. Lions, tigers, and pumas, by contrast, do not eat plants, and are, in consequence, tertiary macro-captors of free energy. Predatory microorganisms tend to be either secondary captors living off plants or tertiary captors feasting on animals. Usually, they do not attack both plants and animals.

This growing specialization in capturing certain types of free energy is an interesting topic by itself that deserves further investigation. In doing so, a great many species found themselves on evolutionary tracks from which there was no escape or return but only a road ahead. This phenomenon is known as path dependency. The general rule appears to be that the more complex organisms become, the more likely it is that they will experience a growing path dependency on their evolutionary track. Yet for lack of space, this important theme will not further be explored in this account.

When did the first tertiary captors emerge, large and small? This is unknown to me. But obviously, their emergence required the presence of sufficient numbers of secondary captors on which they could feed. And when did species emerge that were both secondary and tertiary capture of free energy? Is that perhaps an older survival strategy? And if so, did fully tertiary captors emerge from such species? Furthermore, what term would we use for the microbes and viruses that began to feed on tertiary captors? Would they perhaps be called quaternary captors of free energy? Within the context of this chapter, it is impossible to explore all those questions in any detail. But for a more profound enquiry, answering them is of vital importance.

Among the Cambrian species, the increase of complexity differed. Water-dwelling photosynthetic organisms such as sea weeds remained fairly simple, compared to many of their more recent family members on land. Why would that be? Apparently, adding further complexity to their survival strategies did not offer a clear advantage in the struggle for life, because there was enough water to live in, while its buoyancy counteracted gravity. In consequence, many such floating species did not need any solid stems, trunks, or branches to support their solar arrays.

By contrast, a most important portion of the survival strategy of sea weeds was, and is, to spread out in the water to capture sufficient sunlight, while trying to make sure not to be eaten by other species. And because adding complexity always has an energy price tag attached to it, any organisms that may have evolved forms of greater complexity that did not confer any survival advantages would have obtained a survival disadvantage, and would have been outcompeted by less complex species. Yet floating around entails the risk that such species might be carried by the waters to places where the getting was less good. The first plants that rooted to sea floors may have developed this survival strategy for countering that risk.

During the Cambrian period, novel forms of cooperation may have emerged both within and among species, known as symbiotic relations, among plants as well as animals. Such evolutionary patterns had begun to emerge much earlier in time, perhaps with the first lichen, which consist of combinations of algae and cyanobacteria that are helping each other. Yet during the Cambrian, many more symbiotic relations within and among species may have emerged between microbes, plants, and animals.

Today, for instance, humans and other animals have a great many symbiotic relationships with microorganisms that populate our skin and intestines. It appears to be unknown when such interdependencies first appeared. The emergence of such processes signaled novel types of survival strategies, in which secondary and tertiary captors began to extract free energy and matter from other life-forms without killing them, while also performing beneficial tasks for those organisms.

The rapidly changing and diversifying food-capturing pyramid in the waters may also have influenced the emerging food-capturing pyramid on land, if it already existed, again mostly perhaps through wind-born microorganisms. If so, those developments would have led to continuous changes in the composition and dynamics of the food-capturing pyramid as a whole. But again, for lack of empirical evidence this remains speculative.

THE IMPORTANCE OF EMERGING BRAINS CONNECTED TO SENSORS AND LIMBS AS A NOVEL SURVIVAL STRATEGY

The emergence of complex animals was an important novel aspect of the Cambrian period. Still living in the waters, many of them sported sensors, most notably eyes, that were connected to brains that could direct steerable limbs and bodies. While forming images of the external world inside their brains based on the input of their sensors, this allowed such species to purposefully move around to capture their free energy and matter more efficiently, while seeking to escape the dangers that might threaten their survival. This survival strategy involved more bodily complexity, and was, therefore, expensive. Yet for a great many animals the benefits must have outweighed the costs, because in a great many variations, this body plan is still around. As usual, those innovations led to adaptive radiations, which produced an increasingly diverse fauna living in the waters.

How did this novel innovation emerge? Based on pioneering insights by Hubert Reeves, in Chapter 5 of my book *Big History and the Future of Humanity* a model is presented of how this might have emerged. In summary, this might have gone as follows. As soon as organisms, large or small, began to develop one single primitive sensor, they began to obtain inputs from their external environment that could potentially be useful for improving their survival chances. Our pepper plants, for instance, tended to move their leaves toward the Sun. Apparently, they had not only sensors to detect that but also a mechanism to turn this input into action, with the result that their capture of solar free energy was increased. Yet they did not have any brains that would direct such movements. Why not?

Because pepper plants are tied to the soil by their roots, they are stationary. In consequence, they cannot go anywhere. And even if pepper plants could move around, it would be rather costly to do so while carrying their stems and leaves, much like why today's electric cars do not carry large solar arrays for powering them, because that would be too bulky. As a result, innovations such as eyes, ears, brains, and limbs would yield few, if any, survival advantages to plants. Doing so would instead lead to a survival disadvantage, because the costs would outweigh the benefits. That immediately explains is why our pepper plants could not purposefully move around the apartment in search of solar energy. Yet clearly, their current survival strategy has been sufficiently effective for continuing their existence as a species until the present day.

By contrast, some microorganisms, both prokaryotes and eukaryotes, evolved sensors connected to little steerable limbs, with the aid of which they could move toward food or away from threats. Most of their sensors detect chemical signals, which may therefore rank among the oldest sensors. It appears unknown when, and how, the first sensors emerged. It would not be surprising if the more evolved sensors among the Cambrian species were further developments of earlier microbial bio-engineering.

Unlike the Sun, which is fairly easy to detect by sensors, animals encountered the problem that their sources of free energy, plants and other animals, did not emit similarly easily detectable signals. And while earliest predators may have lived in places where their prey was abundant, so that they would not have needed to make much of an effort in finding it, as soon as their targets became scarce, the process of natural selection would have favored those predatory species who came up with innovations to survive in such leaner circumstances.

A major evolutionary solution to that problem was to purposely go and find such more abundant places. For such microorganisms and animals, it therefore conveyed a clear survival advantage to evolve interconnected sensors, brains, and steerable limbs that could help them go to where the getting was good, even though those organs were costly to maintain. As soon as that happened, adaptive radiations would have produced increasingly diverse numbers of such species, first in the water, later on land, and, more recently, also in the air. As argued above, only secondary captors could do so, because they did not need to carry large solar arrays.

The emergence of species with increasingly large brains that were becoming better at forming images of both the outside world and themselves signaled the emergence of consciousness. Furthermore, it allowed some of them to produce overviews of their direct environment, while also paying attention to details. In addition, those developments made foresight possible: reflecting on what the future might bring as well as adapting one's behavior to such perceptions. All of that must have been very rudimentary in the beginning, yet for some species like ours, those possibilities appear almost limitless today.

Thanks to the fact that brains began to run on software – a process that is still not well understood – having brains also opened up the novel possibility of learning from experiences. As a result, for adapting to the circumstances they found themselves in, increasingly such life-forms were no longer entirely constrained by the process of natural selection, because they could also begin to rely on cultural learning.

Those early brains may have been mostly hardwired with few possibilities for adaptation, if any, much like the early computers that were used in the Apollo space program to land men on the moon. Those onboard computers were literally hardwired, and their programs could not easily be updated as a result. For making the Apollo spacecraft computers of the 1960s, patient and competent women literally sowed those programs in the form of webs of electrical cords that could not be changed during the flights. Yet soon after the Apollo program had ended, computers began to run primarily on software.

Animal brains may have evolved similarly. Their first brains may have been entirely hardwired. But as soon as they became more complex, the amount of cultural software running on those brains would also have increased. This makes one wonder what percentage of information processing may still be hardwired in animal brains today, including ours, and how much of it works like computer software.

Whatever the case may be, animal species endowed with sensors, brains, and steerable limbs began to possess two learning processes for adapting to changing situations: the biological process of natural selection and the cultural process of changing their brain's software. And as soon as such species began to communicate and share such knowledge among members of their own species by signs and sounds, this facilitated the spread of it among them. It also began to make possible inter-generational learning, at least in principle, in doing so potentially leading to the accumulation of knowledge over succeeding generations.

All of this also opened up the possibility for other species to observe this communication and learn from it, such as happens today in a great many situations. For instance, while birds may spot the danger of an approaching predator and may start making warning noises, such signals can also be picked up by other animals, including humans.

In sum: brainy species equipped with sensors and steerable limbs, and endowed with the first forms of culturally-mediated communication, became engaged in increasingly-complex social and ecological interactions. In such situations, the process of natural selection would have rewarded more and better sensors, brain power, and communication as long as the costs did not outweigh the benefits.

As we will see below, such developments proceeded much further on land than in the oceans. Apparently, the circumstances on *terra firma* favored such changes far more than those in the waters, especially perhaps communication. Yet much later in time, some of those smarter land-dwelling species would return to the seas, in doing so introducing those novel capacities also in aqueous environments. And very recently in the biosphere's history, this led to the emergence of one single species, human beings, who, by going to all those places, transformed the biosphere in ways no other species had done before.

To be sure, the emergence of animals with versatile brains should not be considered a goal in biological evolution. It is simply a survival strategy that has worked sufficiently well until today, one of the great many survival strategies within the food-capturing pyramid that also have worked sufficiently well. However, the very brainy human survival strategy may not continue to be successful for much longer. But now we are again jumping ahead in time in our story.

THE INTENSIFICATION OF THE BIOLOGICAL ARMS RACE

The emergence of brainy animals led to an intensifying arms race. As we saw earlier, the phenomenon of arms races among species may have begun much earlier in time, namely after the emergence of the first secondary captors. But we know very little about that, if at all, because by comparison, such arms races would have been tiny in scale, while they happened so long ago that they would have left few traces in the fossil record, if any. Yet by contrast, the fossils of some Cambrian animal species show clear evidence of such an arms race, most notably solid body armor and the first beaks.

But not all novel animals made such things to the same extent. The innovation of brainy animals led to adaptive radiations and a resulting diversity of survival strategies, not unlike what has happened in military history. While some species produced heavy armor, others relied on evasion tactics through higher speed and increased agility. Also living together in larger numbers became part of such a survival strategy. All of the latter is what fishes have been doing since those times, while feeding on large numbers of small plants and animals.

Also squids and similar animals operate without body armor, while they have evolved an additional form of defense that consists of dark fluids that can be ejected to make themselves less visible. More in general, camouflage, deception, and hiding became part and parcel of the more flexible survival strategies, while showing poisonous colors (in doing so seeking to make themselves less attractive) also fall into this category. To avoid being eaten, many primary captors would have begun to defend themselves by producing such poisons, while microorganisms may have done similar things. Others may have done none of that, relying instead on their large numbers and rapid reproduction rates for surviving the struggle for life.

Over the course of time, all complex brainy species evolved heads that contained brains, on the outside of which important sensors became mounted, first of all eyes, which were always situated close to the food-intake area. Those heads usually became connected to their bodies by flexible necks. As a result, those sensors could scan larger portions of the surrounding environment and thus find free energy and matter more easily, while they also became better at detecting dangers to their survival. It is unknown to me to what extent also the Cambrian plants may have evolved defensive traits, perhaps poisons, that would have made they less attractive to the secondary captors that preyed on them.

HOW DID THE BIOSPHERE CHANGE DURING THE CAMBRIAN PERIOD?

What may have been the biospheric effects of the evolving food-capturing pyramid during the Cambrian era? Not a single study known to me addresses this question. As a result, what follows here should again be seen as a preliminary attempt at outlining such possible changes.

In general, those effects must have been considerable, most notably in the oceans and the air, by putting ever more atoms and molecules in unusual places while storing increasing amounts of free energy in organic matter, dead or alive. Doing all of that must further have modified the biosphere, including an increase and intensification of the existing complex cif loops influenced by life. And because some brainy animals were becoming conscious of themselves and their surroundings, this also heralded the beginning of the biosphere becoming conscious of itself, leading to today's human attempts to understand our position and history within our planetary and cosmic surroundings.

The emergence of complex animals introduced a great many novel intentional activities within the biosphere, the effects of which may hardly be known, if at all. By purposefully moving around, by digging, and perhaps by engaging in the first construction efforts, those animals increasingly began to shape their living environments, and, in consequence, the biosphere, including in ways such as Darwin's earthworms were doing in more recent times. And because plants became rooted to sea floors, they would have begun influencing those areas, while they may also have begun to act as keystone species.

All of that must have caused changes in the biospheric complex cif loops, most notably those involving carbon, oxygen, hydrogen, nitrogen, and phosphorus. In addition, complex life also began to increasingly influence other complex cif loops such as the ones that involved calcium, iron, and sulfur, for making bones, blood, and body protection. A few less prominent complex cif loops that became more intense included the increasing numbers of spore elements that were used in life's increasingly intricate biochemistry.

Not only larger organisms began to make constructions out of calcium such as bones, but also smaller life-forms, including microorganisms, started using this chemical element, mostly for making shells to defend themselves. This included the tiny *Foraminifera*, which began to occur in great numbers in the oceans. Today, such microorganisms live, for instance, in the warm shallow waters surrounding south Florida. After such organism die, their shells sediment down to the sea floors. This can produce large accumulations. Later in time, in many places limestone rocks would form out of such sediments, some of which were uplifted to above sea level, such as the famous white limestone cliffs near Dover in southern England. Such ideas were already proposed by the French naturalist Jean Baptiste Lamarck (1744–1818) in his almost unknown book *Hydrogeologie* (1802). Although Lamarck has suffered from a bad press because of his mostly incorrect theory of how biological evolution worked, the French scholar also proposed great novel insights.

The accumulation of calcium carbonate in ocean sediments and the subsequent formation of limestone rocks contributed to keeping calcium out of the water and carbon as well as oxygen out of the air and the water, both for longer periods of time. Yet as soon as such rocks became exposed to erosion, those chemical elements returned to the waters and the air. This was the beginning of life creating major carbon, calcium, and oxygen sinks as part of their associated complex cif loops. In doing so, life began to extract even more carbon dioxide from the atmosphere, which contributed to a further cooling of the biosphere. While tiny calcium carbonate shells accumulated on the sea floor in large numbers, the less numerous animal bones have produced far fewer fossil remains.

In his excellent book *The Story of Earth: The First 4.5 Billion Years, from Stardust to Living Planet* (2012), the US mineralogist and astrobiologist Robert Hazen (1948–) argued that by doing all of that, complex animal life began to add increasing numbers of novel minerals to the biosphere, which made it increasingly complex. Furthermore, as mentioned in James Lovelock's *Gaia: The Practical Science of Planetary Medicine* (2000), marine algae began to emit the gas dimethyl sulfide gas in considerable quantities. This gas stimulates cloud formation and the associated rainfall, which are all part of the complex cif loop related to water. This represents only one example of the ways in which life increasingly began to influence the climate, and, in consequence, the biosphere.

What happened to the production of entropy by life as a result of the Second Law of Thermodynamics? More life means more order, so less disorder (entropy). Yet to satisfy the Second Law, the production and maintenance of all this greater complexity must have required the production of even more disorder. Where did that entropy go? In all likelihood, most of it was emitted into the cosmos in the form of infrared radiation, as it still is today. That did not leave any material evidence.

With the exception of coal, oil, gas, and limestones deposits, today only a few bio-molecules made by life in those times can be found. This means that processes of molecular relooping would have begun to occur, in which life, erosion, and plate tectonics would jointly have played a role. Apparently, from that time the biosphere began to evolve effective relooping mechanisms for most of those bio-molecules. Many of those complex cif loops may have emerged earlier in time. Yet starting from the Cambrian period, they would have become more diverse and intense.

How much would the total amount of living biomass have increased as a result of the Cambrian explosion of life-forms? Studies assessing this topic are unknown to me. As a result, it is impossible to estimate this, including the related question of to what extent those changes would have moved the biosphere further out of thermodynamical equilibrium than it would otherwise have been.

While the Cambrian era was relatively warm, starting from around 485 million years ago a cold period set in, perhaps occasioned by life. This would have eliminated many species, while it perhaps especially affected the more complex life-forms. Yet over the course of time, life bounced back, as it always does, driven by the Fourth Law and the accompanying process of natural selection. Interestingly, the Cambrian period produced the last major innovations of survival strategies that occurred in the waters. All further major innovations and the resulting biospheric changes would happen on land.

THE FOOD-CAPTURING PYRAMID MOVES ONTO LAND

Between 470 and 460 million years ago, the first complex plants and animals would have entered the land. More likely than not, they had been preceded by primary micro-captors, and, possibly, also by secondary micro-captors. The period ever since that time represents no more than about 12% of the biosphere's history.

Seen from a general point of view, many aspects of the invasion of the landmasses by complex life would have been a repetition of the Cambrian explosion of life-forms in the waters, during which complex multicellular life was added to the food-capturing pyramid. Now, this also began to happen on land, with the difference that many important biological innovations had earlier emerged in the oceans. Yet the newcomers could not just step on land and expect to do well, because the terrestrial circumstances were quite different from where they came from.

The major biological innovations that accompanied this transition dealt with the need to survive circumstances in which a ready supply of water was not assured, while the effects of gravity were no longer counteracted by water's buoyancy. That led to the emergence of spacesuits, or rather 'landsuits,' to prevent their bodies from losing too much water while protecting them against both solar radiation and the rapid changes in temperatures between day and night as well as during the different seasons. Furthermore, animals needed to develop lungs to capture the needed free oxygen from the air.

The lack of buoyancy on land counteracting gravity led to a great many innovations. That is how, for instance, many plants acquired upright solid stems supporting their leaves in an effort to capture sufficient solar free energy. This was part of a growing competition among such species for this resource. As usual, those innovations led to adaptive radiations, which over the course of time, honed by natural selection, led to the enormous variety of plants and their associated survival strategies that can be observed today.

For animals, the lack of buoyancy counteracting gravity led to the innovation of solid legs to support their bodies, which also allowed them to move around. Those legs evolved out of fish fins, which were no longer functional on land. Furthermore, we may suspect that those novel developments stimlated the already-present micro-organisms as well as any little newcomers to also undergo adaptations.

Yet life on land was not only hardship, but it also presented fresh opportunities, perhaps most notably the abundant free oxygen in the air, which was there thanks to the persistent activities over billions of years by the first captors of solar free energy in the oceans. This allowed the new landlubbers to develop a greater bodily complexity. Further novel innovations included sensors such as ears that could detect rapid changes in air pressure, which made hearing possible. Also noses further developed, which began to detect the increasing variety of gases that were being inhaled.

Furthermore, organs emerged that began to use expelled air for communication. They were usually part of the food intake openings, which also began to serve for breathing air. By expelling air in more refined ways, and, in doing so, create intricate air pressure patterns, increasingly-complex messages could be sent that were captured by the emerging ears. Apparently, having one single opening that could intake both food and air facilitated the further evolution of that organ so that it could also be used for effective communication. Noses did not have such a dual function, and, in consequence, never became major organs for sending intricate audio messages.

Those forms of audio communication could also evolve because, in terms of expended free energy, the purposeful propagation of sounds in the air is far easier and cheaper than within the waters. This immediately explains why fishes do not talk to each other. Yet they may use other forms of communication, for instance, signaling simple messages with their flexible bodies and steerable limbs that could be captured by eyes. All those forms of communication have continued to exist both in the water and on land.

As a result, the lower atmosphere began to resonate with intentional sounds made by life. Many types of unintentional sounds produced by living beings had already begun to exist in the waters during the Cambrian era, if not earlier. But presumably, most of those sounds would have been much fainter. On land, thanks to the characteristics of the air including the winds, all those sounds became much more noticeable. After intelligent life returned to the waters, the emerging whales, dolphins, and the like, took sound making back into the oceans, this led to remarkable long-distance communication, especially among whales, as well as echolocation among dolphins and the like to find their prey by making clicking sounds and listening to the returning echos. And during their history, humans have increasingly been making all kinds of sounds other than talking to each other, especially perhaps during work, warfare and festivities, and increasingly so in urban settings. This is another rich biospheric theme that deserves further attention, but which, for lack of space, will not further be explored in this account. And again, we have been jumping ahead in time in our story.

The effects of regularly-occurring biospheric changes such as the rhythm of day and night and the changing seasons are much stronger felt on land than in the oceans. Most notably, this is the case with temperature changes, because watery environments tend to dampen such effects. For all complex land-living organisms, this more variable situation put a premium on foresight, even among those without brains such as our pepper plants, in their case by making flowers at the right time as described in Chapter 2. It is unknown to me how many forms of foresight may also have developed among microorganisms that were living on land. But perhaps the formation of spores to survive the lean periods could count as such an example.

But surely, having complex brains opened up the possibility of much greater foresight, of imagining future situations and preparing one's behavior accordingly. This capacity may first have emerged in the waters. Yet it has evolved much further on land, especially among more complex animals, leading to what humans, in principle, are capable of doing today, including a much greater control over many aspects of the biosphere than any preceding species had accomplished.

Like before, this short overview should not be seen in any way as an exhaustive treatment of all the changes that happened when life began to invade the landmasses. Such an attempt would put us again on a long Cape of Good Hope trajectory with the prospect of not reaching our destination within a reasonable period of time.

BIOSPHERIC CHANGES CAUSED BY LIFE CONQUERING THE LAND

Seen from general point of view, life moving on land implied a considerable expansion of the planetary area where solar energy could be harvested, and, to some extent, also geothermal free energy. This was again the Fourth Law in action, with enormous biospheric consequences.

How did the biosphere change as a result of life moving on land? To answer that question, let us first examine some of Earth's geological changes. Also in those times, wind and water erosion would have worn down mountains, as they do today. But after life began to conquer the land, it began to influence those processes. As the Dutch geophysiologist Peter Westbroek explained in his book *Life as a Geological Force* (1992), microbial life such as lichen enhances the erosion of rocks by 'eating' them, by dissolving rocks with the aid of acids to extract the needed minerals. But plants and microorganisms can also keep the soil together, by forming mats, or with their roots. In doing so, they counteract erosion, while their roots tend toward jointly maximizing the intake of water and nutrients, in doing so transforming the soil structure, much like our pepper plants had been doing.

So, life's actions can both stimulate and weaken erosion. Those dual effects led to changing global transport and sedimentation patterns of the particles that were liberated by erosion to wherever they would go: on land, in lakes, rivers, seas, and oceans. By covering ever greater portions of the landmasses, life's influence on erosion patterns intensified. All of this implied a greater influence of life on the 'rock cycle,' in doing so, as Westbroek argued, turning it into an increasingly-complex cif loop influenced by life.

Also the ocean floors increasingly contained life or its remnants. Over the course of time, some of those materials were subducted underneath the continents by the process of plate tectonics. This carried the remnants of life and their effects ever deeper inside Earth. Yet at other times and in other places, those remnants were elevated again to our planet's surface. It is unknown how deep inside Earth life's influence has reached over the eons and what the resulting effects may have been. But because more of our planet's interior became affected by complex cif loops influenced by life, the biosphere expanded in size.

And what about the water loop? Because most plants need a great deal of water to survive and prosper, they increasingly began to influence the biospheric water loop as well, not only its run-off from the land, including ponds, lakes, brooks, and rivers, but also, and perhaps more importantly, its evaporation, and, in consequence, cloud formation and rainfall patterns. This produced large local, regional, and global effects all around the planet, which we are only beginning to understand.

Within this academic field of research, the Russian scientists Anastassia Makarieva and Victor G. Gorshkov (1935–2019) have been doing pioneering work with their 'biotic pump' theory.[5] Their first research was focused on forests that influence the water loop, including evaporation. In doing so, forests cause air pressure differences over large distances that affect air flows and rainfall patterns. Obviously, not only

[5] For an overview of their work, see: https://www.bioticregulation.ru/index.php.

forests influence those water loops, but all of life does. Many more such local and regional loops may exist involving both the landmasses and the oceans. Life moving on land represented, therefore, an intensification of life influencing those major biospheric complex cif loops.

While the biospheric effects of green life moving on land have been extraordinarily large, much of the scholarly and popular attention has instead been focused on the secondary and tertiary captors that made similar moves, probably because they look more like us than plants. Yet what would those animals have had for breakfast, lunch, and dinner every day? Clearly, it was only after sufficient numbers of primary captors had established themselves on land that animals could follow.

THE PAST 400 MILLION YEARS: AN INCREASINGLY COMPLEX AND CONTINUOUSLY CHANGING FOOD-CAPTURING PYRAMID

What happened to the biosphere during the most recent 400 million years up until the emergence of early humans about seven million years ago? As noted earlier, conquering the land implied that life as a whole began using the entire available planetary surface for capturing solar and geothermal energy. As a result, further increases in harvesting free energy in areas where green life had become well-established could only come by improving the efficiency of those biochemical processes under pressure of natural selection.

Over the course of time and driven by the competition for those resources, that led among the primary captors to a great many innovations followed by adaptive radiations. With ups and downs, this produced an increasingly-diverse free energy base of the food-capturing pyramid. Similar selection processes took place among the secondary, tertiary, and quaternary captors of free energy, leading to a similarly growing diversity within the upper echelons of that pyramid.

As part of their competition for solar free energy, land plants evolved a great many different survival strategies for capturing it, ranging from mosses and lichen to the tallest trees, while they increasingly occupied all the available land areas. As a result, life began to cover our planet's entire surface by vegetation, large and small, wherever possible. In other words, the Fourth Law again in action.

The growing diversity among animals produced species whose brains became larger and more versatile. As mentioned earlier, this allowed some of them to escape the competition for resources on land and return instead to the waters, where some of them became successful predators, including today's dolphins, seals, and whales. In doing so, those newcomers changed the aqueous portion of the food-capturing pyramid and its biospheric effects. And thanks to the buoyancy of the oceans counteracting gravity, some of them could become the largest animals in existence today.

Many smarter animals had become warm blooded, which was important for keeping their brains functioning well in varying ecological circumstances. To minimize energy losses in the often colder waters, those species evolved forms of body insulation such as furs and thick fat layers, both of which became attractive to humans much later in time. In addition, some of those species became large and thick-skinned. This represented a survival strategy of defending themselves against predators, much

Growing big and tall on land, competing to capture solar energy: Sequoia trees in the Calaveras Big Trees State Park, California, USA. (June 1996, Photo by the author)

like elephants and rhinoceros were doing on land. This return to the waters by animals makes one wonder how many land plants may also have recolonized the waters.

During the past 400 million years, for the first time in the biosphere's history terrestrial species took to the air by active and purposeful efforts, to make use of the opportunities that this offered. The first to do so would have been tiny insects. Much later, they were followed by the first birds, which evolved from dinosaurs living on land. Because flying is energetically more expensive than walking, there must have been rewards for doing so. Those advantages may have consisted of more efficiently fleeing other predators as well as the ability to catch food more easily, and in more diverse ways, in doing so harvesting whatever free energy and matter was available in the air, on land, and in the waters.

In the oceans, flying fish offer an example of animals that can take to the air to escape their predators. But such animals have never evolved into fully-fledged birds. And much like how flying fish must return to the water being pulled down by gravity, also insects and birds need to come back to our planet's surface after having exhausted their energy supplies. As a result, in contrast to the oceans and the land, the air has as yet not offered stable living conditions for any species. The ones that stay in the air the longest are probably microorganisms blown around by the winds. But to survive and prosper, also they need to come down eventually. As a result, complex ecosystems in the air have not yet emerged.

As part of those changes, increasing numbers of mutual interdependencies began to emerge among the species, for instance between flowering plants and insects, in which nectar, a concentrated source of free energy and matter, began to be offered by the plants in exchange for spreading their pollen around, which helped them improve their genetic variation through cross-fertilization, and, in doing so, enhance their survival chances. This situation offers only one example of an extraordinarily large and

Growing big and tall on land, competing to capture solar energy: Sequoia trees in the Calaveras Big Trees State Park, California, USA. (June 1996, Photo by the author)

complex theme, which, in consequence, has made the biosphere much more complex over the course of time. And, of course, the arms races among the species also continued to evolve.

Punctuated by larger and smaller extinction waves, all those evolutionary processes have accelerated over the course of time, while they were conditioned by the biosphere as it evolved. All of that has led to an increasingly complex, dynamic, and ever-changing food-capturing pyramid, and, in consequence, to an increasingly complex and dynamic biosphere. In the final analysis, all of that happened as a consequence of the growing competition for free energy.

THE PAST 400 MILLION YEARS: AN INCREASINGLY COMPLEX AND CONTINUOUSLY CHANGING BIOSPHERE

How would the biosphere have changed over the past 400 million years, the most recent 10% of its history? While a great many studies exist that examine portions of more complex life and its living conditions during that period, none are known to me that investigate the biospheric effects of the food-capturing pyramid as a whole. Yet those effects must have been enormous, not least because today it is impossible to find one single area of our planet's surface that is totally devoid of life.

Apparently, during that period life became ever more influential in shaping the biosphere, first of all by increasingly harvesting solar free energy. Capturing geothermal free energy, by contrast, must have become a declining survival strategy, because its release from deep inside our planet's interior kept decreasing. Yet in places where geothermal activity still occurs, both in the oceans and on land, it has kept providing free energy to life.

All of this produced a great many biospheric effects, most notably perhaps extracting more carbon dioxide from the air than ever before while delivering more free oxygen into it. But instead of relooping most, if not all, life's atoms and molecules within the biosphere, most notably plant remains became increasingly buried in large quantities as soon as the circumstances permitted it. This process made the carbon complex cif loop even more complex. Over the course of time, this produced the large coal layers which would, much more recently, be exploited by humans.

Why did the accumulation of coal mostly take place during the geological period known as the Carbonaceous, while during more recent periods this happened much less? Were there perhaps not yet sufficient numbers of secondary captors during the Carbonaceous that could effectively feed on that dead biomass? And if so, did the Carbonaceous era perhaps come to an end after species capable of doing so had emerged? That would have implied a further rising complexity of the carbon complex cif loop.

The additional free oxygen in the air allowed more, and more diverse, animals to live on the land, ranging from tiny insects to the largest dinosaurs. This made the use by life of the available free energy more efficient. The more abundant atmospheric oxygen also stimulated a further growth of the ozone layer, in doing so increasingly protecting life on land against harmful solar radiation.

Which novel complex cif loops influenced by life would have emerged during this period? The most important one may have been the complex cif loop related to the emergence of fires. As the US environmental scientist Stephen Pyne (1949–) emphasized in his book *Fire: A Brief History* (2001), as soon as there was enough combustible material available on land as well as plentiful free oxygen in the air, fires could start raging for the first time in the biosphere's history. In other words, the emergence of fires was a direct consequence of life's unceasing photosynthetic activities.

While the appearance of fires further influenced the already-existing carbon and oxygen complex cif loops, this also led to a novel complex cif loop, as Stephen Pyne explained, namely the one that began to regulate the maximum amount of free oxygen in the air. This works as follows. The more oxygen was pumped into the air by life while accumulating combustible carbon on land, the easier spontaneous fires could start. And the more fires were burning, the more carbon would be returned to the atmosphere. At the same time, this also reduced its free oxygen content. In doing so, a feedback loop emerged that tended to stabilize the upper level of the free oxygen concentration in the atmosphere. Over the course of time, with ups and downs this has led to our current atmospheric composition consisting of 21% of free oxygen.[6]

[6] More information on Stephen Pyne and his work can be found on: http://www.stephenpyne.com/.

Life on land also produced an enormous increase of local and regional ecosystems, much like what had already been happening in the waters. In those ecosystems, increasing numbers of species began to live close together, all making use of the opportunities that such situations offered. In doing so, the species involved all became interdependent, also in ways that differed from capturing free energy and matter from each other. Of course, all those ecosystems were connected to each other, sometimes more tightly, and at other times more loosely.

Wherever green life went on land, secondary and tertiary captors followed, large and small, all doing their own things. All those activities further influenced the biosphere, for instance, as mentioned above, in the ways Darwin's earthworms were plowing the soil in search of free energy and matter; or how our pepper plants kept the soil together with their roots; while all the dead biomass that fell down onto the ground changed it beyond recognition, by decaying as well as by being eaten, in doing so producing a layer of topsoil that would never have existed in such ways without life's influence. Earlier in time, similar effects had also begun happening in the waters. It is far beyond the scope of this book to further elaborate any of that.

The biospheric consequences of life taking flight may also have been considerable, because flying, or drifting in the air, made possible traveling over much larger distances, in doing so spreading specific life-forms and their effects very quickly and more efficiently over larger areas, including spreading pollen and pepper plant seeds, or collecting nectar and other resources. It also offered, at least in principle, those flying species the opportunity to diversify their sources of free energy and matter, in other words, to eat whatever was available in the air, on land, and in the waters. In doing so, all of them, large and small, began to influence the biosphere wherever they went.

For a fuller picture of the biosphere's changes during this period we would also need to systematically include all the geological and cosmic effects as they occurred, not only the changing amounts of the incoming solar radiation and the effects of plate tectonics and volcanism but also the impacts of meteorites. Such a famous celestial impact occurred 66 million years ago in what is now the sea just to the north of the Yucatán area of Mexico. It led to the well-known demise of the dinosaurs, at least on land, and of a great many other species as well, in doing so ushering in the dominance of mammals among the larger complex animals.

Although that meteorite impact very much changed the composition of the food-capturing pyramid and, in consequence, its external effects within the biosphere, seen from a more detached perspective it did not change in any fundamental way the structure of both the food-capturing pyramid or the biosphere similar to what had happened at other times, such as, for instance, after life had learned to capture solar free energy; after the emergence of the first complex life-forms; or what happened during the rise of humanity.

To be sure, the emergence and further evolution of mammals, and of grasses, were important changes in terms of biological evolution, and, therefore, also of the biosphere's composition. But until early humans appeared on the planetary scene, those biological novelties did not change the food-capturing pyramid in any fundamental ways, and, in consequence, they did not change the biosphere either in such ways, although local and regional changes and effects were considerable. But that had been

Descendants of dinosaurs visiting our Amsterdam window sill, July 6, 2021. (Photo by the author)

part and parcel of the biosphere's history anyway, within which 'one cannot step twice into the same river,' as the ancient Greek philosopher Heraclitus formulated it.

Yet a few other biospheric novelties could also be witnessed. Starting in times unknown, a few complex animals, large and small, had begun to make the first artificial constructions, such as birds and insects building nests; other animals making holes in the ground; and beavers building dams, while some of them also began to use the first rudimentary tools, most notably perhaps sticks. A few insect species began to build more complex homesteads, such as ant hills and beehives, which were impressive achievements, especially considering their body size. All of that points to an improved intelligence and the related capacity for greater cooperation.

Remarkably, some ants began to cultivate fungus inside their homes, in doing so engaging in the first agriculture. Yet however often this behavior has been cited in the academic literature as an example of animals engaging in agriculture well before humans started doing so, in contrast to such human efforts, ant agriculture has always been practiced on a limited scale; exclusively within their nests; and with a limited diversity of cultivars. In consequence, the biospheric effects of ant agriculture have also remained rather limited, both in size and scope.

Furthermore, while all life-forms concentrate free energy and matter inside their bodies, some land-based animals also began doing so outside their bodies, through storage, with the aim to survive the lean seasons. That was the beginning of another novel biospheric trend that, much later in time, would reach new heights through human actions. It moved the biosphere a little further out of thermodynamic equilibrium than it otherwise would have been. And so did all the other accumulations of free energy and matter by life during this period.

Surely, this summary of biospheric effects during the past 400 million is far from complete. Yet any efforts to further outline this important theme within this account would derail it completely, so that will not be attempted.

CONCLUDING REMARKS

Looking back on the biosphere's history so far, with ups and downs its rising complexity accelerated in the form of a nonlinear process. Yet the decreasing geothermal free energy and the rising incoming solar free energy cannot be seen as direct root causes of this acceleration, although they facilitated it. The conclusion can therefore not be escaped that life must have caused this biospheric acceleration, through the ways in which it has changed the biosphere over the course of time.

The principal root cause was the Fourth Law and its resulting effects: the growing diversity and complexity of the food-capturing pyramid and its resulting biospheric effects, in combination with life's capacity for nonlinear growth. All of that happened under pressure of the process of natural selection and the resulting adaptations of life's survival strategies. That was possible thanks to the adaptability of life's genetic information and the resulting changes in its biochemistry.

This unrelenting pressure of life within the biosphere can clearly be observed today. Today, the high free oxygen and low carbon dioxide concentrations in the air have been brought about by life's photosynthetic activities. Yet even though more recently humans have been adding large amounts of carbon dioxide to the air, its concentration still amounts to only about 0.04% of all the air. That is about the same percentage which is considered the lower limit for successful mining by humans of chemical elements, such as uranium, that are considered economically important. The current low percentage of carbon dioxide in the air provides, therefore, a clear indication of how much competition there has been among the primary captors of solar free energy for obtaining this precious resource. That situation would have stepped up the efficiency of such processes under the pressure of natural selection.

All life-forms became concentrated in places where there was sufficient available free energy and matter. In consequence, other than the global effects through the process of photosynthesis, many of the resulting biospheric effects were first of all concentrated in such areas. Those Goldilocks circumstances were principally conditioned by geological and cosmic aspects. The shapes and sizes of the water and land areas were conditioned by geological effects, while cosmic aspects included Earth's orbit around the Sun and the angle of its axis, both of which determine the amount of the incoming solar energy that falls on a certain area. Also the steady increase of solar energy and the impacts of extraterrestrial materials have played a role.

Over the course of time, all those areas were increasingly influenced by life: by its sheer presence as well as by influencing the related biospheric complex cif loops. The resulting competition for resources led to a dispersal of living organisms to all places with sufficient resources, where they adapted to the prevailing circumstances while changing them through their actions. As a result, life and its biospheric effects spread globally.

The processes of life's first concentration close to rich resources and its subsequent dispersal to less favorable places explain both the presence and diversity of life-forms within the biosphere, as well as all the biospheric complex cif loops influenced by it. Could this unrelenting tendency toward increasing biodiversity therefore perhaps be called another fundamental law of thermodynamics? Or would we regard it as a derivative of the Fourth Law, so, perhaps a Fourth Law A? At the time of writing, I am not sure how to deal with this issue, but currently, a Fourth Law A is my preferred option. To be sure, life-forms also became concentrated, purposefully or not, because of the need for mutual support, including defense against other species. Such processes also tended to concentrate the predatory species that fed on them, again with many biospheric effects.

In this Suez Canal version of our biosphere's history, it has not been possible to do justice to a great many aspects and details. But that was not the goal. In providing this first provisional overview, I hope to have sketched an outline of what needs to be considered for making our biosphere's history better understandable, while it may hopefully also serve as a stepping stone for further research.

In sum: to understand the biosphere's history, we need to examine the entire food-capturing pyramid as it has emerged and changed over time, with all its emerging and evolving survival strategies; the resulting interdependencies among all the species involved; and their biospheric effects in terms of the resulting complex cif loops; and all of this within the context of Earth's geological changes and cosmic influences.

How would the history of our own species fit into this approach? That will be discussed in the next two chapters.

BIBLIOGRAPHY

Allen, Timothy F.H., Tainter, Joseph A. & Hoekstra, Thomas W. 2003. *Supply-Side Sustainability*. New York, Columbia University Press.

Allman, John Morgan. 1999. *Evolving Brains*. New York, W. H. Freeman & Co., Scientific American Library Series, No. 68.

Alvarez, Walter. 2016. *A Most Improbable Journey: A Big History of Our Planet and Ourselves*. New York, W.W. Norton & Company.

Andel, Tjeerd Hendrik van. 1994. *New Views on an Old Planet: A History of Global Change*. Cambridge, Cambridge University Press (1985).

Blankenship. Robert E. 2010. 'Early evolution of photosynthesis.' *Plant Physiology* 154 (434–438). http://www.plantphysiol.org/content/154/2/434.

Bradbury, Ian K. 1998. *The Biosphere*, Second edition. Chichester, John Wiley & Sons.

Broda, E. 1978. *The Evolution of Bioenergetic Processes (revised reprint)*. Oxford & New York, Pergamon Press.

Budyko, Mikhail I. 1986. *The Evolution of the Biosphere*. Dordrecht & Holland, D. Reidel Publishing Company (original edition in Russian 1984).

Carozzi, Albert V. 1964. 'Lamarck's theory of the earth: Hydrogeologie' *Isis*, 55, 3 (September issue), (293–307).

Chaisson, Eric J. 2001. *Cosmic Evolution: The Rise of Complexity in Nature*. Cambridge, MA, Harvard University Press.

Chaisson, Eric J. 2005. 'Follow the energy: The relevance of cosmic evolution for human history.' *Historically Speaking: Bulletin of the Historical Society* 6, 5 (26–28).

Cloud, Preston. 1978. *Cosmos, Earth and Man: A Short History of the Universe*. New Haven & London, Yale University Press.

Cloud, Preston. 1988. *Oasis in Space: Earth History from the Beginning*. New York & London, W.W. Norton & Company.

Conway Morris, Simon. 1998. *The Crucible of Creation: The Burgess Shale and the Rise of Animals*. Oxford, Oxford University Press.

Cook, Earl. 1976. *Man, Energy, Society*. San Francisco, W.H. Freeman & Co.

Crosby, Alfred W. 2006. *Children of the Sun: A History of Humanity's Unappeasable Appetite for Energy*. New York, W. W. Norton & Co.

Darwin, Charles. 1859. *On the Origin of Species by Means of Natural Selection, or the Preservation of Favoured Races in the Struggle for Life*. London, John Murray.

Darwin, Charles. 2001. *Charles Darwin's Beagle Diary Edited by R.D. Keynes*. Cambridge, Cambridge University Press (original manuscript1836).

Debeir, Jean-Claude, Deléage, Jean-Paul & Hémery, Daniel. 1991. *In the Servitude of Power: Energy and Civilization through the Ages*. London & Atlantic Highlands, NJ, Zed Books (1986).

Drury, Stephen. 1999. *Stepping Stones: The Making of Our Home World*. Oxford, Oxford University Press.

Elias, Norbert. 1992. *Time: An Essay*. Oxford, Blackwell.

Emiliani, Cesare. 1995. *Planet Earth: Cosmology, Geology, and the Evolution of Life and Environment*. Cambridge, Cambridge University Press.

Ermann, Dominik & Lubrich, Oliver. 2019. *Alexander Von Humboldt: Das zeichnerische Werk*. Darmstadt, WBG Edition.

Evans, Scott D., Hughes, Ian V., Gehling, James G., Droser, Mary L. 2020. 'Discovery of the oldest bilaterian from the Ediacaran of South Australia.' *PNAS* https://www.pnas.org/cgi/doi/10.1073/pnas.2001045117.

Fox, Ronald F. 1988. *Energy and the Evolution of Life*. New York, W. H. Freeman & Co.

Gould, James L. & Keeton, William T. 1996. *Biological Science*, Sixth edition. New York & London, W. W. Norton & Co.

Gould, Stephen J. & Eldredge, N. 1989. *Wonderful Life: The Burgess Shale and the Nature of History*. New York & London, W. W. Norton & Co.

Graedel, T. E. & Crutzen, Paul J. 1993. *Atmospheric Change: An Earth System Perspective*. New York, W. H. Freeman & Co.

Hamblin, Kenneth W. & Christiansen, Eric H. 2004. *Earth's Dynamic Systems*, Tenth edition. Upper Saddle River, NJ, Pearson Education Inc./Prentice-Hall, Inc.

Hazen, Robert M. 2012. *The Story of Earth: The First 4.5 Billion Years, from Stardust to Living Planet*. New York, Viking.

Humboldt, Alexander von. 1845. *Kosmos: Entwurf einer physischen Weltbeschreibung*. Stuttgart and Augsburg, J. G. Cotta'scher Verlag (1845–1862).

Humboldt, Alexander von. 1850. *Aspects of Nature in Different Lands and Different Climates; with Scientific Elucidations, translated by Mrs. Sabine*. Philadelphia, Lea and Blanchard. (Original Edition in German 1807).

Humboldt, Alexander von. 1997. *Cosmos*, vol. 1, Foundations of Natural History. Baltimore, Johns Hopkins University Press (1858).

Humboldt, Alexander von. 2004. *Ansichten der Natur, mit wissenschaftlichen Erlaüterungen und sechs Farbtafeln, nach Skizzen des Autors*. Frankfurt am Main, Eichborn Verlag (1807).

Knoll, Andrew H., Canfield, Donald E. & Konhauser, Kurt O. (eds.). 2012. *Fundamentals of Geobiology*. Oxford, Wiley-Blackwell.

Lamarck, Jean Baptiste. 1802. *Hydrogéologie, Ou. Recherches sur l'influence qu'ont les eaux sur la surface du globe terrestre; sur les causes de l'existence du bassin des mers, de son déplacement et de son transport successif sur les différens points de la surface de ce globe; enfin sur les changemens que les corps vivans exercent sur la nature et l'état de cette surface*. Paris, L'Auteur, Agasse, Maillard.

Lane, Nick. 2016. *The Vital Question: Why Is Life The Way It Is?* London, Profile Books.

Lovelock, James E. 1987. *Gaia: A New Look at Life on Earth*. Oxford & New York, Oxford University Press (1979).

Lovelock, James. 1995. *The Ages of Gaia: A Biography of Our Living Earth*. New York, W. W. Norton & Company.

Lovelock, James. 2000. *Gaia: The Practical Science of Planetary Medicine*. Oxford & New York, Oxford University Press (1991).

Lunine, Jonathan I. 1999. *Earth: Evolution of a Habitable World*. Cambridge, Cambridge University Press.

MacDougal, J. Doug. 1996. *A Short History of Planet Earth: Mountains, Mammals, Ice, and Fire*. New York, John Wiley & Sons.

Makarieva, A. M. & Gorshkov, V. G. 2006. 'Biotic pump of atmospheric moisture as driver of the hydrological cycle on land.' *Hydrology and Earth System Sciences* Discussion 3 (2621–2673).

Makarieva, Anastassia M., Gorshkov, Victor G., Li, Bai-Lian, Chown, Steven L., Reich, Peter B. & Gavrilov, Valery M. 2008. 'Mean mass-specific metabolic rates are strikinglysimilar cross life's major domains: Evidence for life's metabolic optimum.' *PNAS* 105, 44 (16994–9).

Margulis, Lynn & Sagan, Dorion. 1995. *What is Life? (Foreword by Niles Eldredge)*. London, Weidenfeld & Nicholson.

Margulis, Lynn & Sagan, Dorion. 1997. *Microcosmos: Four Billion Years of Microbial Evolution*. Berkeley & Los Angeles, CA, University of California Press (1986).

Margulis, Lynn & Olendzendski, Lorraine. 1992. *Environmental Evolution: Effects of the Origin and Evolution of Life on Planet Earth*. Cambridge, MA, MIT Press.

Martínez-Frías, Jesús. 2020. *Geología Planetaria y Habitabilidad: Luna, Marte y Asteroides*. Online lecture on November 6, 2020. Burgos, Spain, Asociación Geocientífica de Burgos: https://www.youtube.com/watch?v=5JLkz2xsf3Q&feature=youtu.be.

McNeill, William H. 1976. *Plagues and Peoples*. Garden City, NY, Anchor Press/ Doubleday.

McNeill, William H. 1986. *Mythistory and Other Essays*. Chicago & London, University of Chicago Press.

McSween Jr., Harry Y. 1997. *Fanfare for Earth: The Origin of Our Planet and Life*. New York, St. Martin's Press.

Niele, Frank. 2005. *Energy: Engine of Evolution*. Amsterdam, Elsevier/Shell Global Solutions.

Odum, Howard T. 1971. *Environment, Power and Society*. New York, John Wiley & Sons, Inc.

Odum, Howard T. & Odum, Elizabeth C. 1981. *Energy Basis for Man and Nature*, Second Edition. New York, McGraw-Hill Book Company.

Priem, Harry N. A. 1993. *Aarde en Leven: Het Leven in relatie tot zijn planetaire omgeving / Earth and Life: Life in Relation to its Planetary Environment*. Dordrecht, Boston, MA & London, Wolters Kluwer Academic Publishers.

Priem, Harry N. A. 1997. *Aarde: Een planetaire visie [Earth: A planetary view]*. Assen, Van Gorcum.

Pyne, Stephen J. 2001. *Fire: A Brief History*. London, The British Museum Press.

Raup, David M. 1993. *Extinction: Bad Genes or Bad Luck?* Oxford & New York, Oxford University Press.

Reeves, Hubert. 1991. *The Hour of Our Delight: Cosmic Evolution, Order, and Complexity*. New York, W. H. Freeman & Co.

Smil, Vaclav. 1991. *General Energetics: Energy in the Biosphere and Civilization*. New York, John Wiley & Sons.

Smil, Vaclav. 1994. *Energy in World History*. Boulder, CO, Westview Press.

Smil, Vaclav. 1999. *Energies: An Illustrated Guide to the Biosphere and Civilization*. Cambridge, MA, MIT Press.

Smil, Vaclav. 2002. *The Earth's Biosphere: Evolution, Dynamics, and Change*. Cambridge, MA, MIT Press.

Sperling, Erik A., et al. 2021. 'A long-term record of early to mid-Paleozoic marine redox change' *Science Advances*, 7, 28 (eabf4382). DOI: 10.1126/sciadv.abf4382.

Spier, Fred. 1994. *Religious Regimes in Peru: Religion and State Development in a Long-Term Perspective and the Effects in the Andean Village of Zurite*. Amsterdam, Amsterdam University Press. Downloadable for free as a pdf at: http://www.bighistory.info/bhi_005_035.htm.

Spier, Fred. 1995. *San Nicolás de Zurite: Religion and Daily Life of an Andean Village in a Changing World*. Amsterdam, VU University Press. Downloadable for free as a pdf at: http://www.bighistory.info/bhi_005_035.htm.

Spier, Fred. 1996. *The Structure of Big History: From the Big Bang until Today*. Amsterdam, Amsterdam University Press.

Spier, Fred. 2010. *Big History and the Future of Humanity*. Oxford, Wiley-Blackwell.

Spier, Fred. 2015. *Big History and the Future of Humanity, Second Edition*. Oxford, Wiley-Blackwell.

Spier, Fred. 2017. *How can we examine chance and necessity in big history?* March 2, http://www.bighistory.info/bhi_005_029.htm.

Spier, Fred. 2018. *Concentration and dilution processes in big history*. January 2, http://www.bighistory.info/bhi_005_036.htm.

Spier, Fred. 2018. 'In pursuit of happiness: A first exploration of morality in big history.' In: May Hawas (ed.) *The Routledge Companion to World Literature and World History*. Oxford, UK, Routledge (207–218).

Tarduno, John A., et al. 2020. 'Paleomagnetism indicates that primary magnetite in zircon records a strong Hadean geodynamo.' *PNAS*. DOI: 10.1073/pnas.1916553117.

Toynbee, Arnold J. 1934–61. *A Study of History*, 12 vols. Oxford, Oxford University Press.

Vernadsky, Vladimir I. 1998. *The Biosphere*. New York, Copernicus Springer Verlag. (1926, Original Version in Russian).

Walker, G. 2003. *Snowball Earth: The Story of the Great Global Catastrophe That Spawned Life as We Know It*. New York, Random House.

Ward, Peter D. & Brownlee, Donald. 2004. *Rare Earth: Why Complex Life is Uncommon in the Universe*. New York, Copernicus Books (2000).

Westbroek, Peter. 1992. *Life as a Geological Force: Dynamics of the Earth*. New York & London, W. W. Norton & Co.

Westbroek, Peter. 2009. *Terre! Menaces et espoir*. Paris, Éditions du Seuil.

White, Richard Allen, et al. 2018. 'Viral communities of shark bay modern stromatolites.' *Frontiers in Microbiology*, 13 June, https://www.frontiersin.org/articles/10.3389/fmicb.2018.01223/full.

Wicander, Reed & Monroe, James S. 1993. *Historical Geology: Evolution of the Earth and Life through Time*. Minneapolis & St. Paul, MN, West Publishing Company.

Zalasiewicz, Jan & Williams, Mark. 2012. *The Goldilocks Planet: The Four Billion Year Story of Earth's Climate*. Oxford, Oxford University Press.

6 What Are Humanity's General Effects within the Biosphere's History?

Anthropology is the study of humankind, especially of **Homo sapiens**, the biological species to which we humans belong. It is the study of how our species evolved from more primitive organisms. It is also the study of how our species developed a mode of communication known as language, and a mode of social life known as culture. It is the study of how culture evolved and diversified. And finally, it is the study of how culture, people, and nature interact wherever human beings are found.

Marvin Harris, *Culture, People, Nature: An Introduction to General Anthropology* (1975, Introduction, p. 1).

The biosphere is as much a **creation of the sun** as a result of terrestrial processes. Ancient religious intuitions that considered terrestrial creatures, especially man, to be **children of the sun** were far nearer the truth than is thought by those who see earthly beings simply as ephemeral creations arising from blind and accidental interplay of matter and forces.

Vladimir Vernadsky, *The Biosphere* (1998), original edition in Russian (1928, p. 44).

From the perspective of natural science, both prehistoric human evolution and the course of history may be seen fundamentally as the quest for controlling greater energy stores and flows.

Vaclav Smil, *Energy in World History* (1994, p. 1).

A FIFTH LAW OF THERMODYNAMICS?

How can we understand the history of humanity within the biosphere's history? For doing so, we need to examine our species' past as being part of the food-capturing pyramid. This history can be summarized as follows. Starting from around seven million years ago, step-by-step early humanity succeeded in escaping from a great many nonlinear pressures within the food-capturing pyramid. They did so by learning to avoid becoming food for other species while becoming more efficient in capturing the free energy and matter that had been accumulated in the life-forms they fed on.

The escape from being eaten by others has been a long and difficult road. Because humans could see the large predators that they feared, this made it easier deal with them. Yet until recently, the tiny microorganisms that also preyed on our species

DOI: 10.1201/9781003275350-6

were invisible to human eyes, and thus unknown to them. This made it much harder to control them. Today, many of those little predators are under temporary human control, but not all of them. As the Covid-19 pandemic has shown, such tiny predators can still spread quickly in nonlinear ways while feeding on humans whenever the circumstances are sufficiently favorable.

While seeking to escape those pressures, over the course of time human have increasingly been capturing free energy. Many authors have pointed this out, including Eric Chaisson, Earl Cook, Alfred Crosby, Dietrich Droste, John & William McNeill, Frank Niele, Howard T. Odum, Vaclav Smil, and Leslie White, while interpreting this trend as a major driving force of human history, if not its principal driving force. To be sure, individual human beings, or groups of humans, have not always done so. But jointly, with ups and downs, this has been a dominant trend in human affairs.

A major novel phase in this process took off some 12,000 years ago, when our species began to domesticate plants and animals. In doing so, they began to favor the species that they consumed or used otherwise, while marginalizing or completely eliminating all the rest. From then on, humans began to restructure the food-capturing pyramid to their own needs and desires. No single species had ever done so yet in the biosphere's history.

Because agricultural fields are less biodiverse than undomesticated ecosystems, the emergence of agriculture and animal husbandry led to increasingly large and simplified ecosystems under human control, with corresponding biospheric effects. Furthermore, their partial escape from the food-capturing pyramid has allowed humans to multiply far beyond their reproduction levels, while the planet was not getting any bigger. This led to humanity's expansion and presence all around the globe. As a result, today humanity is reaching the resource limits that make possible its modern existence. Such a trend was already noticed by the UK scholar Thomas Robert Malthus (1766–1834) in his famous treatise *An Essay on the Principle of Populations* (1798), which served as a source of inspiration to both Charles Darwin and Alfred Russel Wallace.

As part of those efforts, humans have constructed, and used, an increasing variety of artificial complexity, ranging from simple holes in the ground to the International Space Station. This development can be interpreted as a form of intensive economic growth, which means that per capita, more wealth is accumulated. No other species had done that yet in the biosphere's history, at least not on such a large scale and with such a remarkable diversity. Doing all of that required humanity to jointly strive toward maximizing the capture of all the available free energy.

Let us quickly survey important phases of that process. Like all larger animals, also early humans began their biospheric career by eating living or recently deceased biomass, both plants and animals. Later in time, they began to use biomass for stoking fires. No other species had done that yet. Starting some 12,000 years ago, humans began to rely on their muscle power as well as on animal traction for practicing agriculture while doing all the other things that they began doing.

As part of humanity's unappeasable appetite for free energy, as Alfred Crosby called it, perhaps from as early as 10,000 years ago, our species began to construct devices for capturing free energy present in wind and water flows, such as sailing boats and, more recently, water- and windmills. For making that work wherever the

Capturing inanimate free energy: grain windmill 't Nupke in 1955 next to the house where I was born in Geldrop, the Netherlands. (From: home movie X1a, original in possession of the author)

available free energy gradients to be harvested cannot be concentrated, large free-energy-capturing surfaces are required. That is why the blades and sails of windmills and sailing ships tend to be large. The free energy in water gradients utilized for driving watermills, by contrast, can to some extent be concentrated, namely by making artificial ponds, which explains why water wheels often tended to be smaller.

In doing all of that, humans began to capture forms of inorganic free energy. Those changes signaled, therefore, the beginning of humanity seeking to capture the entire range of the available free energy, while jointly striving toward maximizing such efforts. That was a unique development in the biosphere's history. No other species had done so before, not even remotely closely.

Much more recently, our species began to burn the concentrated free energy available in coal, oil, and natural gas. All those efforts were intensified after humans systematically began utilizing this free energy for powering machines to make things, at increasingly larger and more diverse scales, all with correspondingly increasing biospheric effects. Because those fuels were the fossilized remains of living organisms that had existed much earlier in time, harvesting them turned humans in a unique form of secondary and tertiary macro-captors. With the exception of a few micro-captors such as *Pseudomonas*, no macro-captors had ever lived off fossilized free energy.

As an aside, while studying biochemistry in the 1970s I had learned about *Pseudomonas*' preference for oily substances. If one wanted to find them, one needed to look no further than the soil near a gas station, while thanks to the activities of those invisible little creatures we could keep driving. By eating the oil that cars dropped onto the roads, they made them less slippery. All of that constitutes interesting complex cif loops as a result of human action, doesn't it?

Let us return to our species' unappeasable appetite for free energy. Constructing dammed lakes and using the free energy stored in them for generating electricity has

Capturing inanimate free energy: the Genneper watermill, used for various purposes, located near my parents' house in the city of Eindhoven, the Netherlands. (Undated postcard, probably from the 1950s, by N.V. Jos-Pe, Arnhem, original in possession of the author)

been another way in which humans turned into primary captors of free energy. This was, in fact, a technically more advanced way of using watermills. Also the recent capturing of solar, geothermal, and nuclear free energy turned our species further into primary captors of free energy. Again, no other species had done any of that yet in the biosphere's history.

Seen from the perspective of the biosphere's history, all of this has happened within a very short period of time, only some seven million years, while most of it was achieved within only 12,000 years, which represents the most recent 0.0003% of the biosphere's history. Those extraordinarily-rapid nonlinear changes effectuated by one single species justify, in my opinion, to call this virtually-unstoppable joint human drive toward maximizing the capture of all the available free energy the Fifth Law of Thermodynamics.

To be sure, all those human efforts were also part of the Fourth Law: life jointly striving toward maximizing the capture of free energy. But unlike any other species in the biosphere's history, our species has jointly, and relentlessly, been seeking to capture all the available forms of free energy. The Fifth Law can therefore be seen to kick in as soon as the members of one single species jointly begin to strive toward maximizing the capture of all the available free energy. This unequaled human drive has been noticed for millennia. It is, for instance, eloquently expressed in the biblical account of creation as: 'let them have dominion over the fish of the sea, and over the fowl of the air, and over the cattle, and over all the earth.'[1]

[1] Genesis 1:26, King James Version. https://www.kingjamesbibleonline.org/Genesis-Chapter-1/.

The Fifth Law in action: a dairy farm north of Amsterdam, April 16, 2019. (Photo by the author)

Human history can, therefore, be seen as an emergent effect of the history of life, as the British astrophysicist, and big history pioneer, Ian Crawford formulated it in reaction to my lecture at Birkbeck University, London, UK, in September of 2019. It is as if suddenly a big bubble formed out of the food-capturing pyramid consisting of one single species, which subsequently rapidly began to transform and dominate this pyramid according to its own wishes and desires, while seeking to capture ever greater portions of the available free energy, in doing so transforming the biosphere. None of that had happened yet in the biosphere's history.

THE EMERGENCE OF A HUMAN FOOD-CAPTURING PYRAMID

As part of that process, humans began to form the first stratified societies, within which some people no longer harvested their free matter and energy from the surrounding nature but instead from other people, as most of us do today in technically-advanced societies. This led to the emergence of a human food-capturing pyramid. It was based on the free energy and matter captured by farmers and herders. Through force and trade, that accumulated free energy became accessible to other people. This development was not entirely new. Especially ants and bees present similar examples. But the human food-capturing pyramid has developed far beyond anything that those little creatures had achieved.

While the tendency of forming such human food-capturing pyramids had not been totally absent in earlier human history, it clearly took off around 6,000 years ago with the emergence of the first early states. For most of their earlier existence, humans had lived in more egalitarian societies. Within such situations, a certain degree of social stratification resulting from power and prestige would also have existed, simply because humans evolved as social animals. But it is assumed that most, if not all the people of those earlier societies were engaged in the daily task of finding, or producing, the required free energy and matter.

Much like what had happened to the natural food-capturing pyramid as a whole, the emergence of the human food-capturing pyramid drastically stepped up the efficiency of humanity's free energy capture and usage, most notably through the division of labor. This produced people with specialized skills performing all those tasks. This development mirrored the human version of the emergence and diversification of the secondary and tertiary captors within the natural food-capturing pyramid.

In today's societies, capturing free energy and matter from other humans against their will is legally defined as 'theft' or 'robbery,' unless done by people who are legally allowed to do so known as tax collectors. The definitions of theft and robbery would have emerged as soon as the concept of 'possessions' emerged, perhaps first of all access to food and land. During the emergence of states, the public pronouncement of such customs came into being as part of governmental efforts to control them.

Seen from a more general perspective, appropriating free energy and matter captured by other individuals against their will is part and parcel of the history of all secondary and tertiary captors. In fact, the entire food pyramid has been permeated by it ever since those species first emerged. And interestingly, the human harvesting of free energy and matter from domesticated plants and animals is hardly ever seen in such ways either, if at all. Apparently, the words 'theft' and 'robbery' only refer to engaging in such activities within the same species as soon as it happens against the prevailing rules.

THE FIFTH LAW COUNTERACTING THE FOURTH LAW?

The Fifth Law of Thermodynamics began to counteract the effects of the Fourth Law related to undomesticated nature, much like how the secondary captors of free energy began to counteract the efforts of the primary captors. Yet through agriculture, those human efforts happened on a far larger scale, namely by drastically reducing the amount of the free energy that was captured by undomesticated life, while at the same time simplifying the food-capturing pyramid in ways no other species had done before. This would have led to a decrease of the joint capture of free energy by all life combined. Yet by harvesting fossil energy starting from the industrial revolution, the total free energy captured by all of life may have increased again, if perhaps only temporarily.

How could we assess all of that? For doing so, we would need to know when, and to what extent, the joint capture of free energy by all of life decreased or increased as a result of human action. Unfortunately, precise numbers appear to be lacking, Yet there can be little or no doubt that until the industrialization of agriculture and animal husbandry with the accompanying large inputs of fossil free energy and fertilizers, agrarian fields and animal pastures did not capture as much solar free energy as they had before they were managed by humans. In that respect, the Fifth Law began to counteract the aspects of the Fourth Law related to undomesticated nature. However, climate changes, especially the end of the last Ice Age and the beginning of the current warm period known as the Holocene, make this even more difficult to assess.

And even though in modern times the capture of solar free energy by corn fields has approached, if not surpassed, the levels of such capture by undomesticated nature, this has only become possible thanks to the enormous input of external free energy in the form of industrial management including fertilizers. If those external inputs were subtracted from the total free energy equation, modern agriculture would not be able to match the capture of solar free energy by undomesticated nature. Yet by tapping into the hitherto unused supplies of fossil and nuclear free energy during the industrial revolution, as well as the free energy present in wind, water, and sunshine, the total joint capture of free energy by all of life, including humans, may actually have gone up during this period as long as those free energy supplies will last.

INTERMEZZO: A FEW EXAMPLES

For many people living today's relatively wealthy lives, the above trends may hardly be noticeable, not least because free energy has often become so cheap and accessible. In addition, compared to human history, let alone the biosphere's history, human life-spans are so short that most people tend to take their current living conditions for granted, assuming that this is the 'normal' situation. Yet for most, if not all, people throughout most of human history, including today's large numbers who still live in far more precarious situations, capturing free energy and matter has been a major daily preoccupation.

A great many documentary sources from the past attest to that unceasing preoccupation. This includes the lively travelogue by Alexander von Humboldt as well as Robert Fitz-Roy's and Charles Darwin's accounts of their voyage of the Beagle. It is also very visible in the fascinating book *La primera vuelta al mundo, 1519–1522* (2018), written by the Spanish naval historian Agustín Rodríguez González (1955–).

This book tells the story of the first voyage around the world that was initially led by the Portuguese mariner Ferdinand de Magelhaes (1480–1521), while it was finished under command of the Spanish sea captain Juan Sebástian de Elcano (1476–1526) after Magellan has perished in battle in what are now the Philippines. This expedition was undertaken by five ships in service of the Habsburg Castilian crown. The only ship that completed the first circumnavigation of the globe in September of 1522 was a little carrack – *nao* in Spanish – aptly named the Victoria, laden with cloves and other spices.

This great adventure was, however, not inspired by attempting to achieve the noble goal of sailing around the world for the first time, even though having done so caused considerable mental effects in Europe and elsewhere. Its main purpose was trying to wrestle the lucrative Moluccan spice trade out of the hands of the Portuguese and make a profit, even though doing so violated the Tordesillas Treaty between Spain and Portugal from 1494 that prohibited Spanish access to those areas. All of that was part of the emergence of worldwide trade and dominion, which represented a novel phase in the biosphere's history as a result of the Fifth Law and its effects.

Over the years that followed, the Spanish were unable to outcompete the Portuguese in the Moluccan spice trade. Yet the first circumnavigation of the globe by Elcano and his crew did succeed in making a profit, even though only one ship out of five returned to Seville while they had also lost most of their crew. As attested by contemporary documents, the Castilian royal court considered that financial profit under royal control its most important achievement.

This was also expressed in the coat of arms that, in 1523, king Charles the First bestowed on captain Elcano. At the bottom, it prominently exhibited a field of cloves, nutmeg, and cinnamon. A castle tower symbolizing the royal house of Castile rose above it, with a soldier's helmet on top of that. And at the very top, the Earth is featured with the lemma: 'Primus circumdedisti me,' 'You first circled me.' A more telling image of the main purpose of that voyage, gaining access to the global spice trade by force under royal control, would be hard to find.

In doing so, the Portuguese and Spanish had begun capturing some of the free energy available in the global flows of wind and water to power their sailing ships, and using it for traveling to lands all around the globe where profits could be made. They could do so thanks to a long series of technical developments stretching over millennia that had turned them into primary captors of the free energy inherent in the winds. Those developments had also included the needed navigational tools and knowledge that allowed them to steer their ships to their desired destinations. All of that allowed Europeans to make such profits, which opened up novel flows of free energy for them, because money is basically a claim on the labor by other people (free energy), including the results of that labor. In other words, this was the Fifth Law again in action.

Another example comes from my Peruvian fieldwork experiences. Until the 1970s, most native Andean farmers had lived feudal lives. They were obliged to work on the large land holdings known as haciendas, which occupied the best agricultural lands. Those were usually the more accessible, flatter, lower situated, and well-watered fields. The native Andean farmers themselves had to make a living by working the more marginal lands situated higher up on the mountain slopes, which were considerably less productive. This led to an enormous joint pressure on the available land to produce as much as possible, in other words, the Fifth Law again in action.

The Peruvian national land reform of the 1970s drastically changed this situation. The haciendas were expropriated and dismantled, while over the ten years or so that followed, the possession of those lands mostly moved into the hands of the native Andean farmers. As a result, they abandoned the higher-situated fields, while concentrating their attention on the better-producing lands that were now accessible to them. This led to a considerable drop in agricultural production, which was a cause for concern for the central government, who needed to feed the growing cities.

This situation resulted in governmental initiatives, including using the aid provided by international 'development' organizations, to improve the agricultural yields by introducing industrial modes of production wherever possible. This included the use of tractors, artificial fertilizers, as well as new varieties of potatoes that yielded more mass per cultivated area. However, the native Andean farmers whom I met did not like those new potatoes. They called them disparagingly *yaku papa*, water potatoes, because even though they were large and more numerous, they contained less starch and more water per average weight. Such potatoes were mostly sold to the nearby city of Cusco, while they themselves kept consuming their traditional produce as much as possible.

Yet the growing Peruvian population over the past forty years has led to an increasing pressure to introduce industrial agrarian practices wherever possible. Once again, this example represented a greater pressure on the land to produce more free energy and matter in line again with the Fifth Law. Yet by 2020, the Andean farmers were complaining that those novel industrial agrarian practices had been damaging their lands, negatively affecting their productivity. Such complaints had already started in the 1980s, when some of them told me that artificial fertilizer was 'burning the land,' in doing so making it less fertile.

As a result, whenever possible, they preferred to use manure for fertilizing their fields, as they had done for many centuries. Yet there are limits to the availability of manure, while over the course of time, rainfall, irrigation, and the harvesting of crops has been leeching from the fields many of the chemical elements needed for growing crops, most notably phosphorus, nitrogen and potassium, much of which as a result has ended up in rivers descending the Andes and connecting to the Amazon River, which transports those materials to the Atlantic Ocean.

The worldwide process of chemical elements required for life being transported from the land to the oceans has been part of the entire biosphere's history, even though through the actions of life, some relooping of those elements has been taking place. But as long as plate tectonics kept uplifting mountains while the erosion of those rocks and soils released sufficient amounts of such chemical elements, life could keep thriving wherever all of that was sufficiently plentiful. Yet human agriculture has been speeding up the loss of those chemical elements much faster than those 'natural' loops can replenish them, which explains why fertilizers are needed for growing sufficient quantities of domesticated plants.

In other words, seen in the longer run, the effects of Fifth Law appear to be undermining themselves by damaging the conditions needed for agriculture. It may well be that the situation in Southern Peru is indicative of the current agricultural changes worldwide. But now we are again jumping ahead in our story.

HOW WOULD OUR ACCOUNT OF THE BIOSPHERE'S HISTORY BE STRUCTURED AFTER THE EMERGENCE OF HUMANS?

So far in our account, the four billion years of biosphere's history have been structured in terms of phases that were defined by the emergence of novel survival strategies of life concerning the capture of free energy. Yet while human history, by contrast, has covered at most a period of about seven million years, it has led to far more incisive biospheric changes during that period than those caused by the rest of living nature. The period of human existence on Earth represents a mere 0.175% of the biosphere's history. In addition, most of humanity's biospheric effects have taken place over the past 12,000 years, which represents no more than about 0.0003% of the biosphere's history, a blinking of the eye, but with profound consequences.

As a result of those sudden large biospheric changes occasioned by humanity, from this point onward our account will no longer be structured by the phases of all of life's novel survival strategies, because, by comparison, all those other species changed far less during that period, while none of them evolved major novelties in their survival strategies related to the capture of free energy and matter.

To be sure, during the past seven million years the biosphere also underwent considerable climatic changes, most notably the ice ages over the past two million years, but also the drying out of East Africa over many millions of years. Furthermore, the process of plate tectonics led to the joining of the Americas, in doing so connecting two large continents, while it also caused mountain formation in South America, Asia, Europe, and elsewhere. Volcanic eruptions and the impacts of celestial objects also caused biospheric changes. Many species needed to adapt to those novel circumstances, including our species, which first emerged on the East African savannas.

Yet none of those ecological changes appears to have led to fundamentally new survival strategies of life concerning the capture of free energy, those of humans excepted. In consequence, our account of the biosphere's history will from now on be structured by the novel phases of human survival strategies regarding their capture of free energy, because that is what increasingly began to produce rapid and profound biospheric changes.

HOW HAVE HUMANS ACHIEVED ALL OF THAT?

Formulating a Fifth Law of Thermodynamics immediately raises the question of how humans could have achieved all of that, as well as why no other living organisms had done so before in the biosphere's history. So, what was novel about those human survival strategies? And how did that cause the sudden acceleration of both human history and the biosphere's history? The major novelty was that those human survival strategies became culturally mediated. In that respect, our species is unique within the biosphere's history. Surely, also other animals have evolved forms of culture. Yet the human cultural achievements dwarf those of all other species.

How did that work? As many authors have noted, through increasingly-effective forms of perception and communication, most notably spoken language, humans have succeeded in interconnecting their increasingly-complex brains. In doing so, humans have shaped and exchanged ever more complex perceptions of reality, including ideas of how to change, or adapt to, the surrounding ecological and social environment. This unequaled human capacity has allowed the communication of mental images with unprecedented size and precision. That has facilitated humans to do all the things that other species had not yet done, including outcompeting all other life-forms in their quest to capture free energy and matter. Spoken language connecting brains did not only lead to the dissemination of mental images, but it also facilitated their accumulation within individual brains. In doing so, increasingly efficient and effective forms of human culture came into being.

How can 'culture' be defined? On page one of his famous book *Primitive Culture* (1871), the British anthropologist Sir Edward Burnett Tylor (1832–1917), by many considered the father of cultural anthropology, described it as follows:

> Culture or Civilization, taken in its wide ethnographic sense, is that complex whole which includes knowledge, belief, art, morals, law, custom, and any other capabilities and habits acquired by man as a member of society.

This definition still appears valid. Yet today, we would not speak that easily anymore of 'primitive culture' because of its possible pejorative connotations. For Tylor and his peers, however, the term 'primitive' simply meant 'the first' without negative meanings, much like how that term is still used in today's Spanish.

How did human culture come into existence? First of all, as just mentioned, not only humans but also many other animals exhibit forms of learned behavior as part of their survival strategies. In that respect, the early humans living between seven and five million years ago on the East African savannas may not have been much more advanced than other contemporary monkeys, if at all. Yet step by step, and, in all likelihood, entirely unplanned, humans learned to improve their communicative and cultural capabilities to the extent that today, humankind has become the dominant species on the planet.

At an early stage on that evolutionary track, humans learned to hang together in bands that were larger than those of the other great apes, while forging the required degree of solidarity to successfully pursue those social lives. Hanging together was a good source of strength and protection on the African savannas, without which those comparatively little and vulnerable creatures might not have been able to survive. Their major daily preoccupation must have been to find food, shelter, and mates, while rearing the young, including trying to make sure they were not becoming food for others. All of that must have required a certain degree of cooperation mediated by culture.

The perhaps most important effect of culture has been to make its participants more flexible, more adaptive to new situations, because it is much easier and quicker to change software in brains than genes in bodies. This was explained in an excellent way by Marvin Harris on pp. 164–169 of his textbook *Culture, People, Nature: An Introduction to General Anthropology* (1975), which I read for the first time in 1980. In fact, Marvin Harris' lucid explanation was a major reason of why I found his book so attractive. In other words, culturally-mediated information can both change and multiply in nonlinear ways. This has provided our species the capacity for learning, adaptation, and inventing new things in ways that are unique in the biosphere's history.

Because cultural learning is not bound to individual bodies in the same way as genes, the theory of natural selection cannot easily be applied to human history as a general principle. It certainly appears to be the case, as William McNeill explained in *The Rise of the West*, that specific cultural skills can become so dominant that all of those who come into contact with people possessing superior skills face the choice of either also acquiring them or become marginalized, if not go entirely extinct. The inventions of agriculture and of industrial production based on fossil fuels offer examples of such dominant skills.

But for successfully adapting to such situations, people do not have to wait for spontaneous mutations to occur in the genes of succeeding generations. The only thing they need to do is to change their brains' software. That is a major difference between genetic and cultural change. The latter allows, in principle, all humans to become successful in adopting new dominant skills regardless of their earlier skills. In trying to do so, the social and ecological situations in which such people find themselves may be conducive to such changes, or they may hold them back, if not entirely block them.

In sum: because of humanity's increasing cultural capability for capturing free energy and use it effectively, our species has become very different from all the other animals that had hitherto existed within the biosphere. As a result, humans would begin to control ever larger portions of the biosphere, at least for the time being.

WHAT DETERMINES THE VELOCITY OF CULTURAL CHANGE?

The aspects that influence the velocity of cultural change are manifold. But a few of them appear to stand out. They include the numbers of humans; how close they live to each other; how many connections they share; how much they compete with each other; and the effectiveness of their communication, including sharing common languages and the state of the communication technology.

Yet while doing all of that, human societies have always had to preserve a balance of maintaining the traditions that allowed them to survive until the present day and adopting novel knowledge that may help them survive in the future. During the Paleolithic, cultural change was slow, presumably because there were few people living in small communities that were only tenuously interconnected. In those situations, preserving the tradition would have had the upper hand, because the group's survival depended on it.

But as soon as the ancient folk began living closer together, especially after agriculture and animal husbandry had begun to take off in the Neolithic while human numbers, densities, contacts, and diversifying survival skills, had all been increasing, cultural change began to accelerate. And because all of that went hand-in-hand with further human population growth, cultural change has kept accelerating ever since.

In doing so, cultural learning quickly took over the process of natural selection as humanity's main adaptive mechanism. From time to time, societies may have repressed innovations, but never for long, seen from the perspective of human history as a whole, let alone the biosphere's history.

TWO MORE BIASES

In many accounts of human history, two rather common biases can be found. The first bias is the tendency to focus on individuals and individual societies perceived as dominant, which are described as having descended from each other in cultural terms. The well-known histories of western civilizations may serve as such an example. In such accounts, all the rest of humanity as well as all the ecology they lived in are considered, at best, their 'social and ecological environment.' This bias, called Bias # 9 in this book, will be avoided in this account.

Furthermore, in most human history accounts it is very common to focus the attention on the greater complexity of the so-called 'civilizations,' most notably the heroic deeds and cultural achievements of those people, while neglecting the rest of the population, the farmers, slaves, and other workers, even though until very recently such people usually made up about 90% of those societies, while they captured and provided most of the free energy and matter that made those urban-based 'civilizations' possible.

This bias, called Bias # 10 in this book, is similar to Bias # 8: focusing the attention on the larger and more spectacular Cambrian animals while neglecting all the rest of life. Also this bias will be avoided in our account. I became painfully aware of this bias while sharing life with my Andean rural family in the 1980s and early 1990s, and I keep experiencing that pain every time I encounter this bias in history accounts, academic or otherwise.

Let us keep those two biases in mind while considering, in the next chapter, the most recent phase of the biosphere's history: its extraordinarily rapid and profound transformation effectuated by our own species.

BIBLIOGRAPHY

Cook, Earl. 1976. *Man, Energy, Society*. San Francisco, CA, W. H. Freeman & Co.

Crosby, Alfred W. 2006. *Children of the Sun: A History of Humanity's Unappeasable Appetite for Energy*. New York, W. W. Norton & Co.

Droste, Dietrich. 2010. *Energiemangel als Antrieb der Menschheitsgeschichte: Eine energetische Gesellschafts- und Geschichttheorie*. München, Martin Meidenbauer Verlagsbuchhandlung.

García Martínez, Adolfo. 2017. *Alabanza de aldea*. Oviedo & Asturias, KRK ediciones.

Harris, David R. (ed.). 1996. *The Origins and Spread of Agriculture and Pastoralism in Eurasia*. London, UCL Press.

Harris, Marvin. 1975. *Culture, People, Nature: An Introduction to General Anthropology*. New York, Harper & Row.

Landes, David S. 1969. *The Unbound Prometheus: Technological Change and Industrial Development in Western Europe from 1730 to the Present*. Cambridge, Cambridge University Press.

Livi-Bacci, Massimo. 1992. *A Concise History of World Population*. Cambridge, MA & Oxford, Blackwell (1989).

Malthus, Thomas. 1798. *An Essay on the Principle of Population, as it Affects the Future Improvement of Society, with Remarks on the Speculations of Mr. Godwin, M. Condorcet, and Other Writers*. London, Printed for J. Johnson, in St. Paul's Church-Yard. 1998, Electronic Scholarly Publishing Project: http://www.esp.org/books/malthus/population/malthus.pdf.

McNeill, John Robert. 2020. *The Webs of Humankind: A World History* (Vol. 1). New York, W. W. Norton & Company.

McNeill, John Robert & Engelke, Peter. 2016. *The Great Acceleration: An Environmental History of the Anthropocene since 1945*. Cambridge, MA, The Belknap Press.

McNeill, John Robert & McNeill, W. H. 2003. *The Human Web: A Bird's Eye View of World History*. New York, W. W. Norton & Co.

McNeill, William H. 1963. *The Rise of the West: A History of the Human Community*. Chicago & London, University of Chicago Press.

McNeill, William H. 1976. *Plagues and Peoples*. Garden City, NY, Anchor Press/ Doubleday.

McNeill, William H. 1984. *The Pursuit of Power: Technology, Armed Force and Society since AD 1000*. Chicago, University of Chicago Press (1982).

Niele, Frank. 2005. *Energy: Engine of Evolution*. Amsterdam, Elsevier/Shell Global Solutions.

Odum, Howard T. 1971. *Environment, Power and Society*. New York, John Wiley & Sons, Inc.

Pollard, Sidney. 1992. *Peaceful Conquest: The Industrialization of Europe 1760–1970*. Oxford & New York, Oxford University Press (1981).

Potts, Rick. 1996. *Humanity's Descent: The Consequences of Ecological Instability*. New York, William Morrow & Co.

Ponting, Clive. 1992. *A Green History of the World*. Harmondsworth, Penguin Books.

Roberts, Neil. 1998. *The Holocene: An Environmental History*, Second Edition. Oxford, Blackwell.

Rodríguez González, Agustín Ramón. 2018. *La primera vuelta al mundo, 1519–1522*. Madrid, Edaf.

Smil, Vaclav. 1994. *Energy in World History*. Boulder, CO, Westview Press.

Smith, Bruce D. 1995. *The Emergence of Agriculture*. New York, W. H. Freeman & Co., Scientific American Library.
Spier, Fred. 2015. *Big History and the Future of Humanity, Second Edition*. Oxford, Wiley-Blackwell.
Tylor, Edward Burnett. 1871. *Primitive Culture: Researches into the Development of Mythology, Philosophy, Religion, Art, and Custom, Volume 1*. London, John Murray.
Vélez, Antonio. 2013. *Homo Sapiens: Psychology as Seen from Evolution*. Bogotá, eLibros Editorial.
Vernadsky, Vladimir I. 1998. *The Biosphere*. New York, Copernicus Springer Verlag. (1926, Original Version in Russian).
Vries, Bert de, & Goudsblom, Johan. 2002. *Mappae Mundi: Humans and their Habitats in a Long-Term Socio-Ecological Perspective, Myths, Maps and Models*. Amsterdam, Amsterdam University Press.
White, Leslie A. 1943. 'Energy and the evolution of culture.' *American Anthropologist* 45 (335–356).
White, Leslie A. 1959. *The Evolution of Culture: The Development of Civilization to the Fall of Rome*. New York, McGraw-Hill.
Wolf, Eric R. 1982. *Europe and the People without History*. Berkeley, CA, University of California Press.
Zalasiewicz, Jan & Williams, Mark. 2012. *The Goldilocks Planet: The Four Billion Year Story of Earth's Climate*. Oxford, Oxford University Press.

7 Seven Million Years of Human Influences within the Biosphere's History

The world of humankind constitutes a manifold, a totality of interconnected processes, and inquiries that disassemble this manifold into bits and then fail to reassemble it falsify reality.

Eric Wolf, *Europe and the People without History* (1982, Introduction, p. 3)

All animals depend on other living things for their food, and human beings are no exception. Problems of finding food and the changing ways human communities have done so are familiar enough in economic histories. The problems of avoiding becoming food for other organisms are less familiar, largely because from very early times human beings ceased to have much to fear from large-bodied animal predators like lions or wolves. Nevertheless, one can properly think of most human lives as caught between the microparasitism of disease organisms and the macroparasitism of large-bodied predators, chief among which have been other human beings.

William H. McNeill, *Plagues and Peoples* (1976, pp. 5–6)

WHAT WAS THE BIOSPHERIC SITUATION WHEN THE FIRST EARLY HUMANS EMERGED?

Around seven million years ago, the continents were beginning to move toward the positions that are familiar to us today. But North and South America had not yet joined, while the subcontinent of India had begun pushing against the southeastern portion of Asia, as it still does, giving rise to the Himalaya mountains.

By that time, the biosphere had recovered from the large celestial impact sixty-six million years ago, when an asteroid had slammed into what is now the coastal area of Yucatan, Mexico. This had caused a severe biospheric disruption, including a huge extinction wave in which the land-living dinosaurs had perished. That had made room for mammals to take over their positions within the food-capturing pyramid. The only dinosaurs that survived were birds. In fact, all the modern birds are thought to have descended from earlier winged dinosaurs. Apparently, flying in the air had been a sufficiently good strategy for surviving the onslaught of that impact.

Over the past thirty million years or so, monkeys and great apes had been evolving, including, about seven million years ago, the first early humans. They were all

DOI: 10.1201/9781003275350-7

secondary and tertiary captors of free energy. The period of the most recent seven million years represents 0.175% of our biosphere's history.

Our early ancestors first appeared on the planetary stage in East Africa. As we will see below, if that particular configuration of the continents had been different seven million years ago, early humans might not have emerged at all. Also since that time, the changing shapes and positions of the landmasses have deeply influenced human history. This includes the emergence of the Mediterranean Sea, the Panama land bridge, and the uplift of the Andean and Himalaya mountains, all with accompanying biospheric changes.

In addition, the process of plate tectonics had concentrated in certain areas materials that later became important human resources, such as flint, gold, silver, copper, tin, mercury, iron, coal, oil, nitrogen, and phosphorus, while diluting them in other places (the process of which has received much less attention, if at all). The biospheric distribution of those resources would have large effects on human history, even though many of them were not utilized by our species until very recently. The principle of concentrated natural resources that cause the concentration of the species that make use of them is generally applicable within the biosphere's history.

As part of those geological developments, the global climate kept changing. Around seven million years ago, the average temperatures on Earth would have been higher than today, while ice-covered poles did not yet exist. The evolving shapes and positions of the continents were leading to changes in water and air flows, which influenced temperatures and rainfall patterns, while volcanic eruptions, earthquakes, and celestial impacts also kept occurring. All of that led to changes within the food-capturing pyramid, with complex biospheric feedback effects.

In other words, the emergence of early humans in East Africa took place within a dynamic biosphere, which to a considerable extent was determined by chance effects. Subsequently, humans began expanding and intensifying their influence in all the available biospheric areas while increasingly interconnecting them as Alfred Crosby emphasized in his pioneering books *The Columbian Exchange* (1972) and *Ecological Imperialism* (1986). Also the similarly insightful book *The Human Web* (2003) by John and William McNeill emphasizes those particular ecological settings and the corresponding human reactions. However, general accounts of the biosphere's history after the emergence of early humans do not yet appear to exist. As a result, again we find ourselves in partially unexplored territory.

WAS THERE ANYTHING SPECIAL ABOUT EARLY HUMANS?

Between seven and four million years ago, an intelligent observer from outer space would, in all likelihood, not have noticed anything special about our early ancestors in comparison with the other large apes. And none of those observers would likely have forecast that from those small roaming bands, a world population would emerge consisting of almost eight billion individuals that were to transform the biosphere in unprecedented ways.

Those early human ancestors are known to science as *Australopithecines*, Southern apes, because their first fossil remains were discovered in South Africa. Their earliest representatives began to roam the East African savannas. Those large grassy plains had emerged while Africa was drying out over millions of years. Much like the other contemporary monkeys and apes, the *Australopithecines* would have made a living as gatherers, scavengers, and hunters. In other words, they were secondary and tertiary captors of free energy.

The drying out of East Africa was not a straightforward process. Wetter and dryer periods alternated, with trees advancing during the rainier times while retreating during the drier periods, leading to the more open savanna grasslands. As a result, over many generations our early ancestors found themselves living in such changing landscapes. And, as 'our' Dutch paleoanthropologist John de Vos used to emphasize, because walking upright on savannas offered a more energy-efficient way of moving around, those early humans started doing that. But the *Australopithecines* kept relatively flexible legs that allowed them to move around in the forests as well.

More in general, the survival strategy of early humans was similar to that of monkeys which mostly lived in trees. As part of that, our ancestors had earlier evolved the capabilities needed to move around in three dimensions. This included, most notably, stereoscopic eyes to see depth; agile hands to grip branches; flexible and strong limbs that allowed them to move around in varying circumstances; and a brain able to process all that information and control all those movements. Another important aspect of their survival strategy was living in groups, with all the benefits and costs that this entailed.

In those times many other monkey species existed, large and small, all around the world. Some of them had been moving over large portions of Eurasia, while others had even reached the Americas, perhaps by crossing the Atlantic Ocean while holding on to floating tree trunks. Yet none of them evolved into early humans.

Why not? Part of the answer is to be found in the process of plate tectonics. In 1972, the Dutch ethologist Adriaan Kortlandt (1918–2009) proposed that early humans living on the East African savannas could no longer interbreed with their cousins who stayed in the forests to the west, because they had become separated from them by a series of deep valleys incised by rivers that they could not cross. This area is jointly known as the Rift Valley. As a result, this region acted as an ecological barrier for our early ancestors.

The Rift Valley geography is caused by plate tectonics. Sometime in the future, East Africa is expected to break off the rest of the continent along the lines of this valley and drift into the Indian Ocean. The early humans began to live on the savannas to the east of the Rift Valley. As a result, the monkeys and apes that still found a home in the tropical forests to the west of it could no longer interbreed with the early humans to the east. This ecological separation would have led to a speciation event, the emergence of a new species.

Roaming the savannas with an upright stride freed the hands of those early humans. This allowed them to start doing things with their hands and arms that other animals had not yet done. Their earlier life in the trees, swinging along branches in search of food, had preconditioned them for doing so. Yet for at least one million years after starting to walk upright, the *Australopithecines* hardly did anything spectacular with their hands. They may have swung around thorny branches to frighten off predators, or have thrown sticks and stones for similar reasons. Furthermore, our ancestors may also have used rocks for cracking open things, while they may have constructed rudimentary nests as well. But none of that would have been very special. More likely than not, other great apes were doing similar things.

INTERMEZZO: THE EMERGENCE OF GRASSES

Between forty and twenty million years ago, the perhaps most remarkable novel aspect of the food-capturing pyramid may have been the emergence of novel varieties of grasses. As described by the New Zealand-born US ecological historian Jonathan Markley (1970–) in his illuminating article 'A child said, "What is the grass?": Reflections on the big history of the Poaceae' (2009), those plants had evolved a more efficient biochemistry for capturing solar free energy. As usual, this innovation was followed by an adaptive radiation, leading to an increasing diversity of grasses. As part of those developments, grasses began to occupy ever larger portions of the landmasses, most notably perhaps the drier areas such as savannas, prairies, pusztas, steppes, llanos, and pampas, whatever name was used to characterize those large grassy plains, while some of them also began to thrive in swamps and on mountains above the tree line. All of this represented another example of the Fourth Law in action.

The rapid expansion of the grasses provided increasing amounts of food to the emerging large mammal grazers, which, as a result, also diversified and multiplied. This led to an arms race between grasses and grazers. While the grasses defended themselves against being eaten by making stalks that were more difficult to digest – including by incorporating the chemical element silicon in their stalks – the grazers reacted by evolving stronger teeth as well as larger and more effective intestines, which allowed them to keep feeding on grasses.

Our imaginary extraterrestrial observers would not have seen anything unusual in this nonlinear growth, spread, and increasing diversity of the grasses and animals that fed on them. Those developments also included a similar increase in numbers and diversity of the large predators, birds, and microorganisms that all fed on those novel plants and animals. Furthermore, the less fortunate species that were being outcompeted decreased in numbers. All of that led to a great many biospheric effects.

However, more likely than not, the great apes rarely ate those abundant mature grasses, if at all. They still cannot do so today, because they lack the long digestive tract needed to do so. As a result, the large apes did not immediately profit from those changes in the food-capturing pyramid. Instead, they preferred other foodstuffs such as fruits, fresh leaves, perhaps some roots, as well as small animals. Digesting such foods required smaller intestines, which allowed the monkeys to move around in more agile ways than the large grazers, which needed to support their heavier bodies on four legs. That immediately explains why those animals did not start walking upright while adapting to savanna life.

Much later in time, some of those grasses and grazers would provide humans with a most important package of food resources, including grains such as wheat, maize, and rice, and animals such as cows, goats, sheep, and horses, the availability of which would condition human history to a considerable extent.

WHAT WERE THE BIOSPHERIC EFFECTS OF EARLY HUMANS?

What may have been the first novel human effects within the biosphere, if any? In all likelihood, there were none that differed from those of other similar animals, while because of their small numbers, those effects would have been limited anyway.

Like many other life-forms large and small, early humans tended to concentrate in small survival groups for defense, protection, and attack. They would have made a living in areas that offered sufficiently rich resources, while some of those places would have been surrounded by less plentiful areas. On the East African savannas, watering holes and lakes would have produced such a concentrating effect, as they still have today, especially during the dry season.

In sum: in terms of the biosphere's history, between seven and four million years nothing fundamentally new would have happened as a result of early human action.

EARLY HUMANS BEGIN TO MAKE TOOLS

Around three million years ago and still in East Africa, the upright-walking *Australopithecines* were beginning to make the oldest-remaining stone tools. Making and using the first such rudimentary instruments was not an entirely novel survival strategy. Other animals had been doing so as well. Yet making the first more elaborate stone tools was probably a novel biospheric development.

What did that yield? Probably an improved diet obtained by hunting larger animals, including breaking up their bones to gain access to the marrow. It also allowed those early folk to defend themselves better or otherwise engage in warfare. Making tools implied an investment in future use. This must have stimulated their brain capacity for learning to think ahead.

The improved capability to harvest free energy and matter with the aid of tools came at the cost of maintaining a larger brain that increasingly guzzled free energy and matter. Those costs also included a more elaborate communication among the group members, which would have led to more cultural learning. Apparently, all of that was a sufficiently-good trade-off, energetically speaking. In doing so, the more elaborate tool use signaled the beginning of a novel path dependency in seeking to survive the struggle for life.

By two million years ago, again still in East Africa, a novel species emerged known as *Homo erectus*, upright man, which became increasingly better at doing all those things. A time-span of a million years for such changes to occur may appear very long. Yet it is rather common in biological evolution. Apparently, the changes that turned those *Australopithecines* into *Homo erectus* mostly came as a result of biological changes under pressure of natural selection, and only to a limited extent as a result of cultural learning.

By the time *Homo erectus* emerged, those early people had succeeded in moving up the food-capturing pyramid thanks to their improving biological and cultural capabilities. Yet their numbers remained small, and our imaginary observers from outer space would still not have had any reason to suspect that those animals would develop into masters of the biosphere. Yet the improved capability of *Homo erectus* to make tools by using their stereoscopic eyes, more complex brains, increasingly dexterous hands, and improved forms of communication, heralded the beginning of a major novel trend that has continued to evolve ever since. It represented the emergence of 'culture' as a major adaptive survival strategy, which increasingly began to influence the biosphere.

INTERMEZZO: WHAT ABOUT THE CLIMATE?

Before explaining how our early ancestors further evolved, we need to pay some attention to the changing climate in those times. The history of our planet's climate is well told by a British team of ancient climate experts, the geologist Jan Zalasiewicz and the palaeobiologist Mark Williams, in their coauthored book *The Goldilocks Planet: The Four Billion Year Story of Earth's Climate* (2012). In a very short summary, Earth's climate has changed a great deal over the eons, while it has never been stable for long periods of time. In fact, the current warm period called the Holocene, which started 10,000 years ago after the last Ice Age had ended, has been unusually stable. But also within the Holocene climate changes have occurred, while currently, human activities are contributing to more such changes.

As we saw earlier, the early humans had partially been shaped by climate change. The drying out of East Africa over the past ten million years or so which had created the savannas was such a climate change. In their turn, those changes resulted from geological and cosmological developments. Yet from about 2.6 million years ago, even larger planetary changes began to occur as a result of plate tectonics, with considerable further biospheric consequences. While India kept pressing against South Asia, in doing so further uplifting the Himalayas, North and South America were joining through the Panama land bridge. All of that changed the conditions for life both globally and regionally, not least because in the Americas suddenly many organisms could spread into large regions where they had not lived before. That led to a restructuring of the food-capturing pyramids in those areas.

But there were also biospheric changes with worldwide effects. The further rise of the Himalayas modified wind and weather patterns not only regionally but quite likely also globally. And the closure of the Panama land bridge blocked off major ocean currents while redirecting other ones. All of that, including the further rise of the Rocky Mountains and the Andes, would have led to additional cooling not only in Africa but also in the rest of the world.

As a result, for the first time since the demise of the dinosaurs, substantial polar ice caps emerged that were perennially covered by ice and snow. This signaled not only frigid conditions in the extreme north and south, but also the onset of the ice ages, in which the polar ice caps temporarily expanded toward the equator, only to retreat again after a while. This first happened with a 40,000 years' rhythm, and later with intervals of about 100,000 years.

The colder periods tended to last longer than the warmer ones. Yet the ice ages were not uniformly cold. Within them, remarkable temperature swings took place over relatively short periods of time. All of that is now thought to be mostly conditioned by the so-called Milanković cycles, named as such in honor of the Serbian mathematician Milutin Milanković (1879–1958), who elaborated the idea that changes in Earth's orbit caused by other planets, most notably Jupiter tugging at Earth during their respective orbits around the sun, would cause climate change on our home planet, because all of that changes the amount of sunlight that falls on specific areas during certain periods of time.

In other words, starting from about 2.6 million years ago, the biosphere's climate began to oscillate over longer periods of time. This may sound a little preposterous. Yet if the entire biosphere's history were equal to a period of one hundred years – the lifetime of a very healthy human being – the ice ages would have begun to occur during only the last twenty-six days of that persons' life, while one single ice age would have had a duration of about twenty-four hours. In other words, by looking at the period of the ice ages from the perspective of the biosphere's entire history, it appears justified to say that from 2.6 million years ago, the biosphere's climate began to oscillate.

What would have been the effects of those climatic changes on early humans? According to the US paleontologist Rick Potts, this would have stimulated the further improvement of tool making and the concomitant brain growth, because *Homo erectus* would have reacted to them by improving their tool kit to capture sufficient free energy and matter. Stated in more general terms, the oscillating of the biosphere's climate would have stimulated those early humans to adapt in both genetic and cultural ways. Any success in doing so would have led to better tool making through cultural change as well as to brainier humans through the process of natural selection. In doing so, *Homo erectus* began to exercise a growing influence over the areas they lived in.

HOMO ERECTUS BEGINS TO DOMESTICATE FIRE

The onset of the ice ages may have stimulated *Homo erectus* to domesticate fire. We do not know when that began, because it is very unlikely that remnants of the earliest fire lighted by humans will be found. As so often, the incipient phase of a process is difficult to detect in the archeological record, if at all. But we should not automatically assume as a result that processes only began at the time of the oldest available evidence.

For various reasons, we may suspect that the earliest human fire use would have started as long ago as 1.5 million years. This period would represent the last 0.038% of the biosphere's history. A major reason for thinking so is that around that time, some *Homo erectus* bands began to move out of Africa into Eurasia. In doing so, they must have encountered conditions that were less mild than those on the East African savannas. Especially colder conditions during wintertime would have stimulated those migrants to make fires, and, quite possibly, also to cover themselves with hides.

By that time, our imaginary observers from outer space might have noticed something new. Perhaps not yet during the beginning of that process, because those primordial human fires would have been difficult to distinguish from the great many wild fires that would simultaneously have occurred. But as soon as *Homo erectus* began to spread over Eurasia, especially while going to places where fires would not occur naturally very often, most notably during the colder and wetter periods, their attention might have been drawn to those curious apes who had started doing what no animal had done before, while they could live in such places as long as there was enough combustible material that could be burned, which usually implied that there were also sufficient other food resources available.

The advantages of fire use for early humans would have been considerable, and have often been outlined: access to foodstuffs that cannot be eaten raw without getting sick; predigesting food by cooking or roasting, which might also have made it more tasty; keeping warm; using campfires for defending oneself against predators, while stimulating group solidarity; using fire for offense and defense among their own kind; and for burning landscapes to create grasslands (artificial savannas) that would attract large grazers, which could subsequently be hunted.

Burning the landscape in the central Sudan, September 1979. (Photo by the author)

A nearby Sudanese Dinka village. For such people, burning the land is part of their lives, September 1979. (Photo by the author)

The domestication of fire made humans more powerful. This may have had considerable effects on their mental state, including their ways of perceiving the world and their position within it. But for our account, its most important aspect was that the domestication of fire heralded the beginning of a novel survival strategy within the biosphere's history. Dutch scientist Frank Niele called it the 'pyrocultural energy regime.'

Human fires may have begun to increase in nonlinear ways both in numbers and extension. It may not take much to light a fire, while a tiny flame can have large biospheric effects, most notably by burning the vegetation of entire areas. If so, fire use would have been the first human culturally-mediated nonlinear effect within the biosphere. In consequence, as Johan Goudsblom argued in his book *Fire and Civilization* (1992), handling fire required both self-control and foresight. This placed a premium on larger and better brains, including the culturally-mediated behavior that enabled our ancestors to do all of that.

Like the effects of tool use, also the effects of fire use would have increased over the course of time. In fact, as we will see below, it was the combination of tool and fire use mediated by culture that turned humans into increasingly powerful animals. In consequence, this combination also lies at the root of how our species began to cause increasingly-profound biospheric effects.

In sum: by taming fire, humans were beginning to strengthen their positions within the food-capturing pyramid, while steadily moving toward its pinnacle. Furthermore, they began to change entire biospheric areas by using the free energy accumulated by life within combustible biomass, in ways that no other species had done before. In doing so, early humans acquired a species monopoly over fire use, as Johan Goudsblom argued in *Fire and Civilization*. In other words, the domestication of fire turned those early folk into a unique biospheric species. In our terms, the domestication of fire by early humans heralded therefore the beginning of the effects of the Fifth Law of Thermodynamics.

BIOSPHERIC EFFECTS OF FIRE USE

What may have been novel biospheric effects caused by fire use? First of all, by burning the landscape in nonlinear ways, the joint capture of solar energy by photosynthetic life may have begun to slightly decrease, and, in consequence, also the joint capture of carbon dioxide. Stated in Vernadsky's terms: through fire use humans began to decrease the free energy accumulated by life within the biosphere. This signaled the beginning of a major novel biospheric trend. From then on, the Fifth Law began to counteract the aspects of the Fourth Law related to undomesticated nature, with increasingly profound biospheric effects.

In addition, by burning entire landscapes with the intention of creating or maintaining grasslands populated by large grazers from perhaps as early as one million years ago, *Homo erectus* began to produce the first biospheric areas that were intentionally created, and simplified. In such places, the joint capture of solar energy by photosynthetic life would have decreased, while by feasting on the large grazers, more of that captured solar free energy would have been channeled toward human use. This major novel trend of simplifying biospheric areas for increased human usage has continued to exist ever since.

Similar, but smaller biospheric effects came as a result of other forms of early fire use, such as stoking camp fires, cooking, etc. Those effects were also smaller because there were still relatively few early humans. Because those early people tended to concentrate in places where there was sufficiently free energy and matter available that they sought to capture, also their further biospheric effects were mostly concentrated in those areas.

In sum: by tapping and managing free energy accumulated by life, for the first time in the biosphere's history our species began to influence the complex cif loops related to combustion. From then on, through human action the biosphere began to move slightly more toward thermodynamical equilibrium than it otherwise would have been. In other words, fire use added a novel layer of emergent effects to the biosphere. All those effects were caused by the growing human ability to control the rest of nature by using their increasingly complex brains, eyes, more dexterous hands, and improving communication.

WORLDWIDE MIGRATION AND THE EMERGENCE OF *HOMO SAPIENS*

Between two million years ago and the emergence of agriculture some 12,000 years ago, nothing fundamentally new happened to the biosphere's history as a result of human action. The biosphere's climate kept oscillating, while humans kept adapting to it. As part of that, *Homo erectus* migrated to many places in Africa and Eurasia, possibly motivated to do so by a growing population pressure and the resulting conflicts about access to free energy and matter.

During those travels, those early humans would have adapted both genetically and culturally to their new environments. This led to a growing diversity among them, including the emergence of species such as the famous Neanderthals. While many of them would have become increasingly competent in doing all the things described earlier, they did not introduce any novel features to the biosphere. Yet their numbers were slowly but surely growing, and so did their resulting biospheric effects.

Moving into colder areas would have caused the need for protecting their bodies. This would have led to making the first clothes, probably from hides. However, making such rudimentary clothes may have been less of an urgent necessity than it may appear to modern urban dwellers. As both captain Robert Fitz-Roy and Charles Darwin described, in the 1830s the natives of Tierra del Fuego – the most southern portion of South America – only partially covered themselves with hides, even though the weather conditions were often cold, wet, and windy.

However, they did oil their bodies well. Apparently, doing that together with wearing a sealskin hide provided sufficient bodily protection against the elements. And they always kept a fire going. In fact, it was because of their many fires that Ferdinand Magellan, who led the first European expedition to those waters, had called this area Tierra del Fuego, Land of the Fire. Early humans may have employed similar tricks to stay warm as long as there was enough oil or fat available. By covering themselves in such ways, the native *Fuegians* made use of the hides, fat, and oil that animals such as sea lions – warm-blooded brainy animals that had returned to the waters – had evolved as body protection over long periods of time through natural selection.

During the past two million years, the early human population as a whole remained small, even though it was growing. Why did it not grow faster in nonlinear ways, in doing so leading to much more rapid population growth? Apparently, there were ecological and cultural conditions that stood in the way of doing so. Those circumstances would have included scarce resources; insufficient brain evolution; the associated insufficient cultural skills to make better use of those resources; infectious diseases; and perhaps also warfare and infanticide. This is not the place to further elaborate such speculations. But if so, those limiting conditions would have kept the biospheric effects of those early humans similarly limited.

While the earlier folk were doing all of that, around 300,000 years ago a novel human species evolved in Africa known as *Homo sapiens*, the anatomically modern humans. They were endowed with larger and more versatile brains, while they were also equipped with hands with opposable thumbs. All of that allowed them to engage in an increasing variety of more challenging culturally-mediated activities. In doing so, the anatomically modern humans increasingly began to replace genetic change by cultural change as their major way of adapting to novel situations.

Starting perhaps as early as 100,000 years ago, also some of them began to migrate out of Africa, in doing so reaching all the continents including Australia, and relatively recently also the Americas. Also this migration process would have been conditioned by ecological circumstances, most notably the last Ice Age. Yet the 'humongous' volcanic eruption of Toba around 75,000 years ago would have produced a 'bottleneck' for human existence by reducing their numbers to only a few tens of thousands. But this view is considered controversial. While settling in novel places, the migrants may have contributed to eliminating large secondary captors of free energy such as mammoth, and perhaps some tertiary captors as well, including saber-tooth tigers.

During their migrations, the anatomically modern humans would have encountered earlier humans, which led to some genetic intermixing. But while the anatomically modern humans survived, over the course of time all the others went extinct for reasons that are still not well understood. Whatever may have happened, within the relatively short time-span of 100,000 years, only 0.0025% of the biosphere's history, *Homo sapiens* became the only human species on the planet that survived. Among them, the general trend would have been to jointly move toward maximizing the capture of free energy by hunting, gathering, fishing, and stoking fires, and all of that with growing numbers while utilizing an increasingly-efficient technology. This was, in other words, the Fifth Law in action.

WHAT ABOUT EARLY HUMAN POPULATION GROWTH?

How rapidly would the early human world population have grown? The known numbers are all very uncertain. If the 'bottleneck' of the anatomically modern humans resulting from the Toba volcanic eruption 75,000 years ago indeed happened, it would have led to a total world population of perhaps 10,000 people. Around 15,000 years ago, according to the Italian demographer Massimo Livi-Bacci in his book *A Concise History of World Population* (1999) there would have been some ten million people. If correct, between 75,000 and 15,000 years ago, the total human population would have increased about one thousand times.

If we assume that a generation in those times would have lasted fifteen years on average, this yields a population doubling rate of once per every 400 generations. If correct, the paleolithic human world population would have grown rather slowly, even though their women would, in principle, have had the capacity to bear ten children or more during their lifetime. Yet if we trust the known numbers, the actual paleolithic growth rate was much smaller, and would, in consequence, hardly have been noticeable in those times.

Why didn't paleolithic populations grow more rapidly, even though they had the biological nonlinear capacity do to so? Speculations range from people seeking to keep population numbers low in various ways because of the limited carrying capacity of the land to their presumably much shorter life spans as a result of wars, famine, infectious diseases, and attacks by predators.

Whatever the case may have been, apparently the paleolithic conditions in conjunction with the contemporary human survival skills considerably limited the human opportunities for nonlinear population growth. And because of their small numbers and similarly limited connections, the exchange of ideas and any resulting cultural changes also proceeded slowly. In consequence, also the paleolithic human influences within the biosphere would have been fairly limited up until the emergence of agriculture.

Yet around 15,000 years ago, in some areas gatherer and hunter life may have been approaching its resource limits given the available technology, while the human numbers kept growing. As the US scholar Marc Cohen argued in his book *The Food Crisis in Prehistory: Overpopulation and the Origins of Agriculture* (1977), this may have contributed to the rise of agriculture about 12,000 to 10,000 years ago, so right after the last Ice Age had ended.

AGRICULTURE AND ANIMAL HUSBANDRY: HUMANS RESTRUCTURING THE FOOD-CAPTURING PYRAMID

We have now reached the "blinking of the eye" period in the biosphere's history: its most recent 12,000 years known as the Holocene, which jointly represent no more than 0.0003% of its entire history. From then on, humans began to transform the biosphere in ways no other species had done before by growing crops and holding domesticated animals. At the same time, the biosphere emerged out of the most recent Ice Age, which led to incisive ecological change. Yet such large biospheric changes after the end of an ice age had happened before. But this time, also humans began to change the biosphere, in nonlinear, accelerating, ways (cf. Mottl et al. 2021). As so often, also the rise of agriculture and animal husbandry may have started considerably earlier in time, while those processes began to accelerate some 12,000 years ago.

During the first few thousand years of the Holocene, our imaginary observers from outer space would have observed fewer trees than expected. In their place, monotonous fields of grasses began to proliferate that were managed by an ape-like species, which was also increasingly controlling groups of herd animals. Over the course of time, those novel plants and animals would grow in numbers and diversity, while the undomesticated nature began to decline. Our extraterrestrial observers had never seen such developments before in the biosphere's history.

The rise of agriculture took place in several biospheric areas, and also in different ways. This is not the place to go into further detail. Over the course of time, all those areas expanded and began to converge, at the expense of the undomesticated biosphere. In addition, domesticated plants, animals, and microorganisms were increasingly transported from their areas of origin to wherever they would flourish. All of this signaled the beginning of the increasing homogenization and simplification of the biosphere as a result of human action.

The transition to a life based on agriculture and animal husbandry implied fundamentally novel ways of capturing free energy and matter. It therefore represented a novel human survival strategy. Instead of seeking to capture whatever free energy and matter was available, humans began to restructure the food-capturing pyramid according to their needs and desires, with the aim to harvest sufficient amounts of free energy and matter. The Dutch scientist Frank Niele called this novel way of making a living the 'agrocultural energy regime.'

Traditional agriculture Inca style in Andean Peru near the village of Zurite: using the chakitaclla, foot plow, December 15, 1985. (Photo by the author)

The domestication of plants and animals led to the concentration in large numbers of a limited selection of species within specific areas. In doing so, humans also began to concentrate themselves, in villages, or, in case of cattle herders, in migrating groups that followed the grasses together with their herds. Because plants are stationary captors of free energy, they were easier to control than animals. Yet in contrast to the moving herds, the cultivated plants tied humans to the areas where they were grown. Unless the animal domesticators combined their lifestyle with growing plants, they remained much more mobile.

As a result, by concentrating and growing the desired species, humans did the same to themselves. Furthermore, in doing so humans created good conditions for other secondary and tertiary captors to prey on them and flourish in nonlinear ways. In other words, also such species became more concentrated in agricultural areas. For achieving success in agriculture and animal husbandry, all those predatory animals, plants, microbes, and viruses, needed to be combated.

The domesticated species were mostly primary captors that formed part of the solar energy-powered portion of the food-capturing pyramid. Species living off geothermal free energy hardly ever appear to have been domesticated, if at all. While this may sound obvious, from a biospheric perspective it needs to be mentioned. The principal crops have been grasses, most notably grains, rice, and maize. And the most important domesticated animals were those that fed on those plants, first of all the large grazers, but also chicken, rabbits, and guinea pigs.

The major reason for that selection of plants would have been that grasses are rapid and efficient photo synthetizers within one single growth season, while producing a great many seeds in nonlinear ways. This allows farmers to store those seeds and feed on them until the next harvest, while having enough leftover seeds to plant at the beginning of the next growth season.

Among the plants, grasses may actually be the Fourth Law champions. And grown in large numbers, they have similar nonlinear effects on the animals that feed on them, which can also reproduce in nonlinear ways, although not as fast as grasses. In other words, unknowingly, humans first of all domesticated the Fourth Law champions of the biosphere as well as the animals that captured their free energy and matter from them. In doing so, our species became the Fifth Law champions, because no other species could compete with them in that respect.

Humans did not exclusively cultivate grasses. Other major domesticates included plants with edible roots such as potatoes, cassava, and yams, which also grow and ripen in nonlinear ways within one single growth season. All the rest was extra, including a wide range of beans valued for their high protein content, for instance, as well as olives and sunflowers for their oil. Potatoes do well in temperate zones, where they yield about four times as much free energy and matter as grains per cultivated area. As a result, they often became the preferred crop wherever they could be grown. Cassava and yams are tropical plants that thrive in areas where the domesticated grasses would not do well.

While domesticating animals, humans also looked for nonlinear growth and maturation. As part of those developments, also pigs became a major domesticated species because of their rapid nonlinear growth potential while eating many kinds of leftover plant materials. In recent times, considerable efforts have been made to boost those growth rates even more. And, of course, not only plants but also herds could grow in nonlinear ways as soon as the circumstances permitted it.

Other domesticated plants such as sago trees, fruit trees, herbs, spices, plants for medicinal and coloring uses, wood for fire use and construction, etc., have also been important for human sustenance. But they have never become the major crops that supported large agricultural societies. Also our pepper plants belong to this rest category. They were probably domesticated for adding a certain enjoyment to the food, but not for their calories. This explains why pepper plants are not grown in similarly large quantities as grains and potatoes, row after row, field after field. As a result, the biospheric effects of pepper plants have remained limited.

Over the course of time, forms of longer-term exploitation also came into use, most notably perhaps milking and producing wool. Because milking yields a relatively constant flow of highly-valued free energy and matter, doing so must have occasioned less immediate attention to nonlinear growth. The same applies to hens laying eggs. Also wool, traction, horses riding, etc., were longer-term uses for which rapid nonlinear growth was less important, although never entirely absent. The British archeologist Andrew Sherratt (1946–2006) called those developments the 'secondary products revolution.'

All of that increasingly facilitated human nonlinear growth, at least in principle, depending on the circumstances, while also other life-forms that could exhibit rapid nonlinear growth, such as predatory microorganisms as well as invasive plants and animals, sought to profit from these novel circumstances.

As part of the domestication process, humans have exerted selective pressures on plants to make them grow larger edible portions. This similarly happened to domesticated animals after they had become tame enough to handle. During their initial domestication, people would have selected the weaker, smaller, and more tame animals that could be dominated by humans. But after that was done, the animals under human control could become bigger again, in doing so producing more food and other resources. Grass-fed domesticated animals also began to be used for replacing human muscle power for traction and riding.

Engaging in agriculture introduced a novel diversity among the domesticated species, for providing more varied diets as well as for hedging the risks against inclement weather and infectious diseases. Today's large mono-cultures and the concomitant decrease of biological diversity represent a very recent development, which only became possible with the advent of industrial agriculture.

In doing all of that, humans created for themselves a monopoly on the exercise of legitimate violence with regard to their domesticated plants and animals. This was the first time in the biosphere's history that such a monopoly emerged at such a scale. During the emergence of states thousands of years later, a similar monopoly on the exercise of legitimate violence would also emerge within human societies. This will be elaborated below while discussing that period.

Even though through agriculture and animal husbandry the yields appropriated by humans of free energy and matter per area became considerably higher, producing those calories implied more work. This explains why many people did not want to start practicing agriculture as long as they had a choice. But if they opted out, they were invariably replaced by more powerful agriculturists over the course of time.

Such a process took place, for instance, in Europe starting from around 7,000 years ago, when large portions of that continent were colonized by farmers coming from the Middle East. That migration was part of a global process, of which we are witnessing the final phase today. In doing so, humans took their domesticated plants and animals with them to wherever all of them could thrive.

As part of those developments, the ratio between the consumable harvest and the seeds that needed to be saved for planting would have increased, as the Dutch historian Bernhard Slicher van Bath (1910–2004) analyzed in his well-researched book *The Agrarian History of Western Europe, A.D. 500–1850.* Although those developments occurred much later in time, they represent another clear example of the Fifth Law in action. One would expect that such efforts and the resulting changes would have begun much earlier in the process of agrarianization.

Because agriculture and animal husbandry were more efficient ways of capturing solar energy than gathering and hunting, this became the dominant human survival strategy. For those engaging in it there was no way back, because the growing population needed to be fed while the available land did not increase, unless one moved somewhere else. As a result, starting from 12,000 years ago humanity found itself on an inescapable path dependency toward controlling all the available land within the biosphere by agriculture and animal husbandry while intensifying their production. This led to more efficient ways of capturing solar energy and matter.

This situation explains why land ownership became important in human history. To be sure, also gatherers and hunters would have known territorial claims. But especially in sedentary agrarian societies, within which the people had become tied to the land and its products, land ownership became a major source of wealth. This ownership also implied control over people, if only to make sure that such claims were respected, as well as, later in time, to make other people work those lands for the landlords' benefit.

This short overview cannot possibly do justice to the extraordinary diversity of domesticated plant and animal species that have come under human control over the past 12,000 years and the resulting human societies. But within the context of this account, it may suffice.

THE EMERGENCE OF THE DOMESTICATED FOOD-CAPTURING PYRAMID

During the millennia that followed while human numbers grew and their societies diversified, more, and increasingly diverse, species were domesticated. This led to a decrease of the untamed biodiversity, and thus to a simplification of the biosphere. And because agricultural land captures less solar energy than undomesticated areas, the effect would have been a decrease of the aspects of the Fourth Law related to undomesticated nature while stepping up the effects of the Fifth Law.

The agricultural revolution implied a fundamental restructuring of the food-capturing pyramid as a whole. In doing so, our species began to produce a domesticated outgrowth of that pyramid which, over the course of time, would become increasingly large. But even today, the domesticated portion of the food-capturing pyramid has not yet become fully independent of the larger pyramid, not least because certain undomesticated plants, animals, and microorganisms that profited from agriculture and animal husbandry also became part of the domesticated food-capturing pyramid. Such species include animals such as wolfs feeding on sheep; birds living off grains and other domesticates; mice and rats that feed on a large range of domesticated plants; as well as insects and microorganisms that feed on all of them. In doing so, all those undomesticated organisms also became part of the domesticated food-capturing pyramid.

Why had all those developments not happened earlier in the biosphere's history, and how did out species succeed in pulling off this trick? The answer is simple and obvious, yet with far-reaching consequences. Humanity did so by influencing the food-capturing pyramid with the aid of culturally-mediated behavior that stimulated the growth of desired species while seeking to eliminate the undesired ones.

As Charles Darwin noted in his book *On the Origin of Species*, the domestication process represented therefore the human cultural equivalent of the process of natural selection. And anatomically modern humans were the first larger species in the biosphere's history that was equipped with such talents. It was therefore at this particular period in the biosphere's history that cultural selection began to take over the dominant role of natural selection for determining who were to live and prosper as well as who were marginalized, if not entirely would go extinct.

To be sure, natural selection has kept functioning, as it does today. Yet starting from 12,000 years ago, the process of natural selection has increasingly been influenced by human cultural selection. To some extent, the earlier human use of fire had already produced similar effects. And as part of agriculture, fire practices became very important for clearing the land from unwanted vegetation such as trees and 'invasive' plants as well as for doing all the other things that can be done with the aid of it.

As a result of the emergence of those human culturally-mediated skills, the food-capturing pyramid as a whole began to split up into a growing domesticated portion under human control and a declining undomesticated portion. In recent times, also that undomesticated portion has increasingly come under human control, namely to preserve and protect what is seen as 'wild' or 'untamed' nature. In doing so, also those portions are being domesticated, but in their case not in order to extract from them as much free energy and matter as possible. Over the course of 12,000 years, this has led to a biosphere which is dominated by a domesticated food-capturing pyramid created and maintained by human beings. And all of that has happened in only 0.0003% of the biosphere's history!

As part of those developments, agrarian human societies began to form little internal human food-capturing pyramids. This happened because some people were becoming more powerful than others, which allowed them to extract free energy and matter from other people. Within the earlier gatherer and hunter societies, such incipient social hierarchies would not have been entirely absent. But because the farmers had become tied to the land and their societies, the weaker agricultural people could less easily escape the actions of their dominant fellows. As a result, over the course of time such social hierarchies became more pronounced.

WHAT ABOUT THE RESULTING POPULATION GROWTH?

What happened to the world's population numbers between 12,000 and 6,000 years ago, when the first early states began to emerge? According to data provided by Massimo Livi-Bacci, they would have risen from between one and ten million people 12,000 years ago to between twenty-five and fifty million 6,000 years ago. Those numbers would include the growth of agrarian societies as well as the decline of gatherers and hunters. Surprisingly, even though agrarian societies became much more productive, the total population doubling rate during that period would only have risen from once per 400 to once per 300 generations.

Yet this increase signaled the beginning of what could be called a process of 'super-nonlinear growth,' a type of increase that is also known as super-exponential growth (Hanson 2000). This occurs as soon as the velocity of nonlinear growth begins to accelerate. And even though the human world population numbers during most of its history are rough estimates, the trend appears clear: an increasing velocity of human population growth. This came as a result of humanity's increasing culturally-mediated skills of capturing free energy.

Why did the agricultural human population growth rates not become higher? Perhaps this was partially caused by the extensive growth of those societies, who were spreading around the world instead of engaging in more intensive agricultural growth. In doing so, they may have harvested less produce from one single area than they otherwise might have done, which may have made it more difficult to maintain large families. Furthermore, forms of birth control, violent conflicts, and increasing infectious diseases may all have played a role in keeping the growth rates still relatively low.

During that period, what would have been the numbers and growth rates of the domesticated plant and animals as well as a possible decline of undomesticated species? And how much free energy and matter was harvested from domesticated species during that period? Such numbers appear to be entirely lacking in the scholarly literature. For the past 200 years or so, some of those numbers as well as their ecological impact have been well described by, among others, the British geographers I.G. Simmons, Andrew Goudie, and Neil Roberts. Yet for a good understanding in quantitative terms of humanity's biospheric impact during the entire history of agriculture, also a deeper knowledge of the earlier numbers is required. Again, we encounter a research field that appears open to further investigation.

As mentioned above, at the beginning of the domestication process at most around ten million people would have lived on the globe. Today, there are almost eight billion of us. This represents an increase of eight hundred times. Yet in the meantime, the planet has not become any bigger. This means that the biosphere currently produces eight hundred times as much human food as it did 12,000 years ago. Those numbers alone already sufficiently justify the idea of the Fifth Law of Thermodynamics. And this calculation does not yet take into account the growing amounts of fossil and inorganic free energy harvested by humans in recent times.

And what about the decrease in the harvesting of free energy by the undomesticated portion of the food-capturing pyramid over the past 12,000 years? Again, this is unknown for lack of data. Initially, the warmer and wetter Holocene may have provided increasingly good conditions for undomesticated species, in doing so offsetting any early decline caused by the domestication of plants and animals. Yet the decrease of harvesting free energy by undomesticated nature must have become noticeable after the human numbers and their activities began to grow in nonlinear ways while the climate had stabilized.

WHAT ABOUT STORING FREE ENERGY AND MATTER?

Cultivating plants has rarely automatically yielded continuous supplies of free energy and matter. Much like how, in 2017, we quickly harvested our peppers by the end of the growth season, the availability of such free energy and matter flows produced by life usually occurs in rapid spurts. Yet to sustain their bodies, the humans and their animals needed to capture free energy and matter on a daily basis. As a result of that situation, the introduction of agriculture inevitably led to the need for storing those products for longer periods of time.

This was a novel issue that most gatherers and hunters had not needed to face as long as they could follow the plants and animals on which they depended. Yet as soon as such early folk came to depend on rich plant and animal resources that were not available all year round, such as fish along coastlines, also they experienced a storage problem that needed to be resolved.

While domesticating animals, this was less of an issue as long as the herds could keep moving from one feeding ground to the next, because the animals themselves provided the required storage of free energy and matter. But if they were stationary, a need emerged to store animal fodder. Such a situation introduced the problem of how to protect the food supplies against natural influences such as inclement weather as well as against the predation by other life-forms, humans included.

Storing food is not an exclusively human affair. Some animals do so, too, such as bees and squirrels. In fact, one may wonder whether humans picked up the idea of storing food from such animals. If so, one would expect an overlap between areas in which such animals were already doing that and the first centers of plant domestication. But whatever the case may have been, one would expect that those early farmers were excellent observers of the rest of nature. Many traditional farmers still are, as I experienced in the Peruvian Andean village of Zurite during my cultural anthropological fieldwork in the 1980s and 1990s. As my *compadre* Hilario Cconucuyca once told me while walking across the fields: "we are ever surprised how gringos walk on the land, as if they do not pay attention to anything."

WHICH NOVEL BIOSPHERIC EFFECTS BEGAN TO OCCUR AS A RESULT OF AGRICULTURE?

By increasingly replacing undomesticated primary and secondary captors of free energy with a few selected domesticated species, the undomesticated portion of food-capturing pyramid began to shrink while its domesticated portion increased. As a result, those human activities increasingly began to influence the already-existing complex cif loops, which, as a result, changed both in size, composition, and velocity. All of that began to transform the biosphere.

This may have included an incipient decline of the capture of carbon dioxide by domesticated photosynthetic life. Initially, however, such possible effects would hardly have been noticeable, not least because the simultaneous onset of the warmer Holocene would have provided better living circumstances to all of life, which would have led to a larger joint capture of both solar free energy and carbon dioxide.

A major portion of the biospheric effects came as a result of humans increasingly changing regions covered by an undomesticated climax vegetation into cultivated lands. Seen from an ecological point of view, those lands were disturbed areas within which invasive species thrived. In addition, by plowing those fields their erosion patterns changed. Furthermore, also the increasing construction activities, mostly housing and storage facilities but also tools and other forms of artificial complexity, all began to influence the related biospheric complex cif loops.

It is impossible to go into much further detail here, not least because such an integrated approach still needs to be written, while the needed specific information about this period as a result of human action appears to be scarce, if not entirely lacking. But a few aspects deserve to be mentioned, perhaps most notably the agricultural technologies that were developed to improve crop yields, such as irrigating lands; the leveling of mountain slopes by constructing terraces; the drainage of swamps; the construction of polders; making floating fields such as in Mexico and Peru; crop rotation; as well as using fertilizers.

Let us quickly examine only one of those examples, the utilization of fertilizers for enhancing the agrarian production, beginning with night soil. Relooping all of that back onto agrarian fields produced the first partially human-controlled complex cif loop that involved human and animal excrements. More likely than not, this first happened within their immediate environments, but later also over intermediate distances, especially after cities emerged.

Why would fertilizers be needed in the first place? First of all, harvesting crops and transporting them elsewhere removed from the land the chemical elements needed for further growth, most notably nitrogen, phosphorus, and potassium. In the second place, the disturbed landscapes created by agriculture opened them up to more water and wind erosion. This tended to remove the topsoil that contained the needed chemical elements, most of which would end up in the oceans as a result.

All of that may to some extent be replenished by breaking up the soil, including the resulting microbial action that liberates chemical elements that were locked up within the soil particles. That is why leaving the land fallow for a year or so may restore its fertility to some extent. Also regular flooding may bring in fertile soil that has a similar effect. Yet in most cases, the continuous practice of agriculture has tended to remove more needed chemical elements than those that are replenished naturally. That is why adding fertilizers is necessary for maintaining a sufficient agrarian production. While animal dung and night soil have played major roles as fertilizers in many places, the net balance has often remained considerably more negative for agricultural lands than for areas covered by the undomesticated food-capturing pyramid, also because all those 'wild' organisms jointly tend to keep the soil together.

In reaction to the declining fertility of the soil, fertilizers began to be mined, such as, for instance, the concentrated dried-out bird excrements (*guano*) that had been deposited on rocky islands along the Pacific Coast of South America. The *guano* was first used by indigenous Amerindian societies, and, in the nineteenth century, also by Europeans who began to ship it to the old continent. Those bird excrements had been deposited on those islands in large amounts because nesting was safe there, as no land-dwelling predators could reach those islands, while thanks to the cold Humboldt ocean current, fish food was abundantly available close by. The European-led '*guano* boom' in South America was followed by a similar commercial boom after nitrate-rich saltpeter deposits were discovered in the nearby Atacama desert. This discovery led to a war between Peru and Chile between 1879 and 1884, which was won by Chile.

By the beginning of the twentieth century with the aid of the Haber-Bosch industrial process, Europeans had learned to capture nitrogen from the air and turn it into fertilizer. They were the first animals to do so, but not the first living organisms. The earlier cyanobacteria, and later also the Rhizobium bacteria in collaboration with leguminous plants, had already started doing so much earlier in time. The Haber-Bosch process became possible as soon as there was sufficient free energy and industrial technology available to do so. No longer limited by a lack of nitrogen, from then on humans further improved the opportunities for domesticated plants to grow, and, unintentionally, also for other species. Yet such modern fertilizers are destroying the associations between crops and microorganisms, especially those microbes that had earlier captured nitrogen from the air in exchange for sugars excreted by the plant's roots. But now, we have again jumping ahead in time in our account.

The chemical element phosphorus, by contrast, essential for all of life, cannot be captured from the air, the water, or the land, because it usually occurs in rather diluted ways, if at all. It can only be mined in places where it had already previously been concentrated by geological or biological processes, such as the *guano* by birds. In sum: by using fertilizers, humans increasingly began to further influence the complex cif loops of especially nitrogen, phosphorus, and also potassium.

Other biospheric effects of the agricultural revolution include the use of fire for clearing land, for cooking, warfare, and for making all kinds of things, in doing so further influencing those complex cif loops. By making tools, clothes, dwellings, storage places, the first carts, increasing amounts of luxury items, burial sites, places of worship, and probably much more, humans began to further shape the biospheric areas where they lived according to their own desires. All of that had biospheric consequences, some local, some regional, and some perhaps already global, the latter perhaps most notably through the large-scale burning of the land.

While doing all of that, humans increasingly began to accumulate certain chemical elements considered valuable outside their bodies, most notably gold, silver, and precious stones, in doing so increasingly putting also those chemical elements in places where they otherwise would not have been. Those human efforts were not unique in the biosphere's history. Some other animals had been doing so, too. But the scale and diversity of those human artistic efforts were unprecedented. And to make all those things, the required resources needed to be extracted from places where they had already been sufficiently concentrated by other processes.

To sum up: by increasingly turning the food-capturing pyramid that had evolved over billions of years into a domesticated pyramid within the time-span of only 12,000 years, our species began to dominate the biosphere in a process of nonlinear change, with a velocity as if it had been struck by a sudden natural event such as an asteroid impact or a supernova event. Yet while the effects of such cosmic impacts may be profound, the events themselves are of short duration. They are measured in minutes, or perhaps in hours or days. The current unprecedented biospheric change effectuated by humans is, by contrast, less sudden but perhaps more persistent and pervasive.

EARLY STATES: THE EMERGENCE OF HUMAN FOOD-CAPTURING PYRAMIDS

While agricultural societies grew and advanced, around 6,000 years ago the first human food-capturing pyramids began to emerge. Within those social structures, growing numbers of people began to acquire their required free energy and matter no longer directly from the domesticated nature but instead through other humans. Expressed in more familiar terms, this was the rise of the first early states. In those times, the world population would have consisted of twenty-five to thirty million people. Now we are talking about the most recent 0.00015% of the biosphere's history. The process of early state formation would have been preceded by thousands of years of slowly-growing social inequality, which became more pronounced some 6,000 years ago.

What would our imaginary observers from space have noticed? First of all larger concentrations of people, the first cities, some of which began to sport large artificial structures shaped like pyramids. They were surrounded by agricultural lands dotted by smaller concentrations of people, the villages. Within the cities and villages, more fires would have been burning, while fire use on the land would mostly have been restricted to the dry season when people sought to get rid of unwanted materials including vegetation.

Within those emerging human food-capturing pyramids, the emerging urban elites began to appropriate and use increasing quantities of free energy and matter captured by the farmers in exchange for protection. This represented the emergence of a novel human survival strategy. Social hierarchies and the accompanying prestige – expressed by the production and use of prestige goods – became pronounced aspects of those societies.

To bolster and express their privileged positions, the emerging leaders needed the support of the growing urban middle classes, including the materials made by them, while they began to rely on armed groups to maintain this novel social order. All of that led to an increasing pressure on the farmers to capture and handover as much solar free energy and matter as possible. With ups and downs, states began to grow and spread around the world. Those changes jointly produced increasing effects of the Fifth Law.

The emerging human food-capturing pyramids bore many similarities to both the undomesticated and domesticated food-capturing pyramids. This included attack and defense, theft, robbery, and conquest as well as the potential to grow in nonlinear ways whenever the opportunities presented themselves. The latter is witnessed by the fact that within less than one thousand years after the early states had emerged, the first larger states already took shape.

Thanks to the emergence of elaborate writing, for the first time in history it became possible to trace the lives of individual people known by name. They were usually members of the elites, while, especially in trade, they also belonged to the

State formation in Andean Peru: the central Inca sun temple Qorichancha in Cusco, with the Roman Catholic Dominican church on top of it, July 1996. (Photo by the author)

Large stone walls of the Inca fortress of Sacsayhuaman above Cusco, July 1996. (Photo by the author)

middle classes. All of them needed to keep records to control 'their' free energy and matter flows, and as a result, also to control other people.

Hardly any farmers, by contrast are known by name from that period, if at all, because they did not keep any known written records. Doing so might actually have been dangerous, because it entailed the risk of losing more possessions to the elites.

Instead, more likely than not farmers memorized all of that, including favors given and received. This is still part and parcel of the intricate social networks of traditional rural societies today.

All of that led to Bias # 9 mentioned in the previous chapter: paying a great deal of attention to the elites and other city dwellers while neglecting the farmers, even though they formed by far the largest portion of such societies, usually up to 90%, while they captured most, if not all the free energy that made those states possible.

As a result, agricultural land, and also mines, became coveted possessions and tended to end up in the elites' hands. Up until the industrial revolution, those resources were considered the major source of income for the wealth of nations. To be sure, also commerce was becoming a major source of wealth. But that was always based on what farmers and miners provided. Conflicts over those resources led to a great many wars, much like how people are fighting over oil and other resources today. In those conflicts, farmers usually made up the bulk of the armies.

How did early states emerge? Most notably the US anthropologist Robert Carneiro (1927–2020) saw ecological and social constraints as the main root causes, most notably growing farmer populations that had become tied to the land in fertile river valleys, who could not go anywhere else because those valleys were hemmed in by deserts. Over the course of time, that situation allowed the emerging elites to do their power grab. Furthermore, the sea level rise of about 120m after the end of the last Ice Age led to the inundation of areas that had formerly been inhabited, including the Arabian/Persian Gulf. This added to the ecological constraints and further facilitated early state formation.

INTERMEZZO: WHAT ARE STATE SOCIETIES SEEN FROM A SOCIOLOGICAL POINT OF VIEW?

What is a state? From a sociological point of view, a state is a constellation of interdependent people ruled by elites who control two indispensable monopolies. According to the great German sociologist Max Weber (1864–1920), the main state monopoly is the legitimized use of physical force in the enforcement of its order. In modern states, the police usually take care of that within those societies, while defense and attack is usually the task of armies.

The second state monopoly is, according to Norbert Elias, the right to levy taxes. This monopoly is also essential, because state complexity has a price tag attached to it that needs to be paid. As soon as one of the two monopolies breaks down, the state collapses. Those monopolies did not appear overnight. They emerged as long-term processes. Furthermore, those two monopolies may never be absolute nor entirely uncontested. But for states to exist, they must be sufficiently robust to allow governments to stay in power.

How can the importance of those two monopolies quickly be demonstrated? While teaching the emergence of states as part of my big history courses, I used to ask the students what they would do if suddenly both monopolies disappeared. In other words, from one moment to the next, there would no longer be a government, and no police either who could be called for assistance as soon as something happened that was considered against the law.

As soon as the students began to realize the magnitude of the problem, they saw that such a sudden change would immediately lead to social chaos, within which hardly anything would work anymore in the usual ways, including personal security, shops, money, etc., while physically stronger people might try to dominate and exploit the weaker ones. So, the first thing they would do, was to form alliances with like-minded people while seeking to obtain sufficient arms to defend themselves and get access to the needed resources.

Because most people would do so, this would lead to a situation in which all those newly-emerging tribes would start fighting each other. Much like what has happened many times in human history, over the course of time that would lead to the emergence of new states and the associated pacification of the population, including the re-emergence of both monopolies. This situation immediately explains why, even though states may fall apart after a certain period of time, they tend to reemerge and spread, because they are such stable human configurations.

POPULATION GROWTH IN STATE SOCIETIES

What happened to the total human population numbers during this period up until the beginning of early globalization starting in 1492 CE? They would have risen from twenty-five to fifty million people about 6,000 years ago to some 500 million people about 500 years ago. This would have represented a ten-fold increase or more. Also the velocity of this growth stepped up again, reaching an average doubling rate of once per one hundred generations. This was considerably higher than the doubling rate of 'free' agrarian societies (once per four to three hundred generations), even though the farmers had to pay taxes and were also exploited in many other ways, while urban populations usually did not grow by themselves but rather through migration from rural areas.

Apparently, states offered better survival conditions than the earlier 'free' agrarian societies, even though this included farmer exploitation; increasing infectious diseases; external wars with the resulting loss of life; and all of that while the available arms were becoming more destructive. This suggests a rather unpleasant picture of life within and among those earlier 'free' agrarian societies. Such a view would be very much in line with the ideas about the decline of violence during human history presented by the Russian psychologist Akop Nazaretyan (1948–2019) and the Canadian-American psychologist and linguist Steven Pinker (1954–).

The growing numbers of urban people provided increasingly good circumstances to their predators, mostly microbes and viruses, which as a result could grow in nonlinear ways as soon as the circumstances permitted it. That led to the first large epidemics. Those waves of infectious diseases were also facilitated by the increasing and more rapid connections among urban people over larger distances. Furthermore, the growing amounts of domesticated animals also provided good circumstances for such disease-causing microorganisms, which might subsequently make the jump to human societies, as they do today.

THE INCREASING CAPTURE OF INANIMATE FREE ENERGY WITHIN STATE SOCIETIES

The capture of inanimate free energy by humans had already begun to take off earlier in time within many agrarian societies, perhaps most notably the use of sailing boats. Yet within state societies, such forms of free energy capture considerably increased, in doing so seeking to capture the free energy present in wind and water flows, with the aid of wind- and watermills. Those devices converted the linear wind and water flows of free energy into rotary movements that could power an increasing variety of applications.

Also the comparatively recent invention of gun powder was a novel way of tapping inanimate free energy, at least partially, because the carbon that forms part of it was produced by life. Doing all of that turned humans increasingly also into primary captors of free energy. Furthermore, fossilized carbon was increasingly used for stoking fires. But because that carbon consisted of leftover plant remains, in that respect humans remained secondary captors of free energy.

Wind- and watermills were relatively complex and therefore expensive, while they could rapidly be destroyed by enemies. In consequence, their exploitation only became feasible in areas under relatively stable state control. That explains why 'free' agrarian societies found little or no use for them. Sailing boats, by contrast, could be useful also to smaller societies, not least because they allowed quick attack and escape.

Larger sailing vessels became the prerogative of state societies, because those people could afford the needed investment. Also because such ships could carry relatively large amounts of goods, they became important in trade and conquest. Yet the winds were not always reliable. As a result, large warships propelled by human muscle power long remained an important option as well. All of that represented efforts to replace animal and human muscle power with captured inanimate free energy. As soon as such technologies became sufficiently feasible, they spread to all the areas where they were deemed useful.

In sum: by increasingly bulging out of the undomesticated food-capturing pyramid while restructuring it, the human food-capturing pyramid also began to exploit sources of inanimate free energy. The joint result was a further stepping up of the Fifth Law. This led to increasing biospheric influences, none of which appears to have been described yet in an integrated fashion.

BIOSPHERIC EFFECTS OF STATE SOCIETIES

The accelerating development of state societies speeded up the resulting biospheric changes. This included effects from the growing human numbers; their increasing interactions and interconnections; their growing skills of capturing free energy and matter as well as of fire and tool use. In doing so, humans increasingly influenced rural and urban landscapes, while making more, and more varied, amounts of artificial complexity. As a result, humans increasingly transported atoms and molecules to all kinds of places where they would otherwise not have been, with profound biospheric effects that still need to be described in detail.

Which novel biospheric developments in state societies appear to stand out? Those were, in the first place, the emergence of large agricultural landscapes with growing amounts of domesticated plants and animals, which also included undomesticated species that sought to profit from that. And in the second place, the emergence of cities, with the first large buildings, including the transportation facilities that connected all those areas. In the third place, the increasing mining, and use, of rocks and minerals. All of that began to influence the associated already-existing complex cif loops, such as those of gold, silver, copper, tin, mercury, iron, and lead. As part of those developments, novel substances were created that had not yet existed before, most notably perhaps concrete, fired bricks, and glasses. In sum: step-by-step, humans began to influence growing numbers of chemical elements and their combinations as well as the associated complex cif loops.

Over the course of time, most of the materials concentrated and used by humans inevitably became diluted again. As a result of human fire use, for instance, mercury and lead, with relatively low evaporation temperatures, often became airborn and were displaced as a result over sometimes large distances. Rocks and other building materials would erode and pulverize over the course of time, if they were not destroyed by human action. In doing so, this added to the dust generated by agrarian and construction activities. As a result, the complex cif loops of all those particles and chemical elements were increasingly influenced by humans.

Urban areas began to counteract the Fourth Law even more than agricultural regions, because in cities less harvesting of sunlight took place than in the surrounding countryside. Yet those growing urban human concentrations offered new chances for predatory microorganisms, also because the nearby larger concentrations of domesticated animals offered them good food, as William McNeill described in his book *Plagues and Peoples* (1976). All of that introduced novel human influences in the associated complex cif loops.

In sum: the emergence of state societies led to an intensification of the effects of the Fifth Law and the associated decrease of the effects of the Fourth Law related to undomesticated nature, while humans began to influence more biospheric complex cif loops, and increasingly intensively so. At the same time, the undomesticated food-capturing pyramid further declined, while its domesticated equivalent kept growing, together with the increasing human food-capturing pyramids. Those were all intertwined processes that produced growing and increasingly-complex biospheric feedback effects.

THE GLOBALIZATION OF THE FOOD-CAPTURING PYRAMIDS

Starting from 1492 CE, when Christopher Columbus and his expedition entered the Caribbean area, human societies became globally interconnected in ways that had not yet existed before. By that time, about 500 million people would have been living on our planet. The period from 1492 CE up to the present day represents the most recent 0.000013% of the biosphere's history.

Earlier in time, the large Eurasian-African continent had already become interconnected by human action through overland travel as well as by navigating along its coastlines. To a lesser extent, that had also been the case within the Americas and Oceania. But while until that time the native American population appears to have lived rather isolated from the other major world areas, a few sea voyages along the western Pacific coastline perhaps excepted, Oceania and the Eurasian-African continent had already loosely been interconnected through human efforts.

Those novel European-led global interconnections were effectuated through ocean travel, by using wind and sea currents (both indirect forms of solar free energy). In other words, the fact that humans had become captors of those forms of inanimate free energy made the process of early globalization possible. This went hand-in-hand with the European use of free energy stored in gunpowder for firing bullets and cannon balls at people considered enemies. Furthermore, Eurasian infectious diseases wrought havoc in the Americas, which contributed a great deal to allowing Europeans to fortify their novel presence in that part of the world. This is, of course, a very short summary of what was a very complex globalization process. Many of its aspects have been well described by eminent scholars such as William and John McNeill, Alfred Crosby, Jared Diamond, Dennis Flynn and Arturo Giráldez, as well as by Eric Wolf in his pioneering book *Europe and the People without History* (1982).

Seen from a general perspective, the human species, which had evolved on land, suddenly also began to dominate the seas and oceans. As part of that, humans stepped up their efforts to influence those aquatic environments, most notably through hunting animals that contained desired supplies of free energy and matter, such as many types of fishes, but also other animals such as whales and seals. By moving out into the seven seas, humans began to dominate oceanic world areas in which other brainy and formerly land-locked animals such as whales, dolphins, and seals had preceded them.

Through those human efforts, all the existing food-capturing pyramids became globally interconnected. This led to a massive restructuring of Earth's ecology. This was well described in Alfred Crosby's pioneering books *The Columbian Exchange* (1972) and *Ecological Imperialism* (1986). People, plants, and animals were all moved around the world to places where they could profitably live, at least as seen from a human perspective. They were often unintentionally followed by predatory species and other life-forms that hitched a ride.

Interconnecting and restructuring all the world areas had enormous biospheric consequences, many of which may have been similar to tectonic plates colliding to form a new super-continent, a second Pangea, as the German-born US historian Wolf Schäfer (1942–) called it. It could be argued that the first human globalization wave had already taken place much earlier in time, when anatomically modern humans began migrating to all the continents except Antarctica. That helps to explain why, starting from 1500s CE, humans encountered other humans almost everywhere they went.

Our imaginary observers from space would suddenly have noticed that the oceans were increasingly dotted with sailing ships, and that domesticated plants, animals, humans, their products, technological inventions, as well as human settlements were suddenly appearing in a great many places where they had not been before, and that all of that was rapidly evolving. All those changes resulted from

Plowing with oxen near the village of Zurite, October 1988, a result of the first wave of globalization. (Photo by the author)

humans seeking to capture free energy and matter wherever possible all around the globe. In our terms, this was the Fifth Law again in action, in doing so further counteracting the aspects of the Fourth Law related to undomesticated nature.

What happened to the growing human numbers? Around 1500 CE, there would have been a world population of about 500 million people. As a result of the globalization process, between 1500 and 1750 CE the human world population rose to about 770 million people. This means that its doubling rate suddenly increased to once per thirty generations. Also those data come from Livi-Bacci (1999, p. 31). That increase took place even though the native American population drastically declined as a result of the Eurasian infectious diseases, the wars, and the often harsh colonial exploitation.

The rapidly-improving means of communication increasingly interconnected all those growing human populations. That led to a larger, faster, and more intensive dissemination of cultural information and material products, including domesticated plants and animals of many kinds. This speeded up the cultural learning processes.

BIOSPHERIC EFFECTS OF EARLY GLOBALIZATION

All those changes led to incisive ecological changes which transformed the biosphere increasingly rapidly. While the biospheric trends from earlier times continued to exist, then they acquired global effects. The fact that people and their domesticated plants and animals could all exhibit nonlinear growth made those global biospheric changes especially fast and incisive. And even though all of them were transported around the world in relatively small numbers, this inbuilt capacity for nonlinear growth produced an extraordinary rapid biospheric transformation.

I encountered a telling example of this process during my Peru research. In the 1530s CE, led by the conquistador Francisco Pizarro a small group of Spanish adventurers had succeeded in overthrowing the Inca empire. Before that time, European crops and animals had not yet existed in that area. Yet in January of 1586 CE, only fifty years later, a systematic survey ordered by the Spanish Crown executed all over the conquered Americas yielded the following information about Antapampa. This is the small and flat Andean plain, about 15 km^2 in size, in which the village of Zurite is situated where I performed my research. The Antapampa plain is situated at about 3,300 m above sea level. The report stated, among other things, that:

> It supports 40,000 head of cattle, cows and sheep, pigs and mares. Much grain, barley, maize, potatoes, oca, and quinua is harvested, the latter being food of the Indians. Also many vegetables; while at the present day peach trees are being planted. [...] And on the surrounding mountains a lot of cattle is wandering around.
>
> *Jiménez de la Espada (1965, p. 201)*

All of that ecological (and social) change had been effectuated by Europeans in a mere fifty years, thanks to life's capacity for nonlinear growth. Such processes began to take place all around the world wherever feasible, and have continued to do so ever since. In the Americas, this was made possible as a result of the sudden nonlinear expansion of the Spanish and Portuguese empires. This was copied by other European states wherever the social and ecological situations permitted it.

Part of those developments was the emergence of more and larger mono-cultures under commercial pressures. This implied a further simplification and uniformization of the biosphere. The general effect was, again, putting ever more atoms and molecules in places where they otherwise would not have been, while increasingly influencing growing numbers of complex cif loops.

While all of that was going on, another novel acceleration of the biosphere's history caused by human action began to take place: the industrial revolution.

INTERMEZZO: LEVI JOSEPH SPIER VISITS THE 1904 ST. LOUIS WORLD'S FAIR

> When the Fair is lighted, when all buildings illuminate the plaza with their floodlights or aim them to the sky, when everything on the Pike radiates life and cheers out of sheer joy, when the gondolas are darting across the ponds and underneath the lighted bridges, America, land of the unlimited possibilities that has created a paradise out of a wilderness, and has provided something so majestic to behold that infinitely surpasses everything that had existed until today, then celebrates its triumph as a world empire that has gathered all people to witness its youthful prowess.

With this exuberant description, the Dutch businessman Levi Joseph Spier (1866–1944) summarized his impressions of visiting the 1904 St. Louis World's Fair. He was my paternal grandfather, and his assessment forms part of an exciting travelogue of the 'conducted tour' that he took to the United States in July of 1904. This tour was organized by the Hamburg-America Line steamship company. At that time, also other tour operators, including Thomas Cook of London, offered similar guided tours to the St. Louis World's Fair. This grand industrial exhibition was also known as the Louisiana Purchase Exposition, because it celebrated that about one hundred years earlier, the United States had acquired that enormous stretch of land from France.

My grandfather's as yet unpublished manuscript contains a great many other insightful descriptions of how that portion of the world was changing under the impact of the industrial revolution, including mass migration to the United States on the very same steamship, the Pretoria, on which he was traveling in far more luxurious circumstances. All of that had suddenly become possible as a result of humans tapping into large but hitherto rarely utilized sources of free energy, first mostly coal, and later also oil, while doing unprecedented things with them.

INDUSTRIALIZATION: CAPTURING INANIMATE FREE ENERGY AND USING IT TO DO THINGS

About 250 years ago – a super-fast blinking of the eye in the biosphere's history – humans began to harvest the large supplies of solar free energy that had been accumulated by life much earlier in time, while they utilized it for powering machines. Now we are talking about the last 0.000006% of the biosphere's history. Our imaginary observers from space would have seen the Earth suddenly lighting up by all kinds of fires, accompanied by growing smoke and soot clouds, both on land and in the oceans, while the human numbers and the domesticated species were all rapidly growing.

Constructing steam engines and using them for all kinds of purposes represented another series of unprecedented nonlinear changes in the biosphere's history. All the industrial engines – starting with steam engines, followed by internal combustion engines, steam turbines, and electric motors – as well as their great many applications jointly exemplified forms of increasingly-refined tool and fire use. In doing so, the process of industrialization represented a novel human survival strategy, called the' carbo-cultural energy regime' by Frank Niele. The incipient phase of stoking carbon had earlier emerged in various portions of Eurasia. But initially, coal, and also oil, were not used for powering machines.

In consequence, during the industrial revolution humans further extended their role as secondary captors of free energy. As part of those changes, water, wind, and muscle power were increasingly replaced by fossil-fueled power. In addition, during the twentieth century the increasing industrial exploitation of nuclear, wind, water, and solar free energy by converting them into electrical power turned humans further also into primary captors of free energy. They were the first animals that engaged in such activities. Until that time, other animals could not have done so, because they needed to move around. And that does not work very well, as mentioned earlier, while carrying large sails or solar arrays.

Even though humans could not eat those novel sources of free energy, thanks to their cultural achievements they could use them for powering machines that converted fossil free energy into rotary movements, which, like those of the earlier wind- and watermills, could power a rapidly-increasing variety of applications. But unlike wind- and watermills, steam engines could also power boats and trains. No other animals had done so yet in the biosphere's history.

As explained in Chapter 3, the conversion into useful forms by steam engines of the free energy present in coal could not be achieved with 100% efficiency. That issue led to the science of thermodynamics, and thus also to the theoretical approach used in this account. Seen from a general point of view, the rapidly-increasing human capture of inanimate and inorganic free energy represented a further spectacular advance of the Fifth Law.

As usual in the biosphere's history, those novel inventions led to a series of adaptive radiations, in doing so producing a growing diversity of novel human survival strategies. This led to fast and profound social and ecological global changes. Suddenly, the production of many items could be organized on a much larger scale, namely in factories powered by steam engines, while the same applied to transportation, communication, selling, farming, mining, as well as the emerging tourism and entertainment business.

The novel access to those inanimate free energy flows also allowed increasing numbers of people to live in cities, which, like the factories, began to produce much more smoke, day and night, while they also started to light up at night. None of that would have escaped the attention of our imaginary observers in space. Growing numbers of urbanites began to feel increasingly disconnected from the rest of 'nature,' while they lost much of their knowledge about it. Yet among a few pioneering urban scientists, a novel growing awareness of the importance of the 'natural environment' could also be witnessed. This included coining the term 'biosphere.'

As part of those changes, more free energy began to flow within the human food-capturing pyramid. Much of it accumulated near its apex in the hands of the emerging industrial elites. Over the course of time, also the increasingly wealthy middle classes began to profit from those changes, while the earlier landed elites lost power,

A freight truck leaving the village of Zurite, November 1988, a result of industrialization. (Photo by the author)

because their societal strength depended on what was then becoming a lesser source of free energy: taxation resulting from agriculture.

Today, many world areas have caught up in the industrialization process. Yet enormous local and regional differences have remained. As a result of all those developments, the global human food-capturing pyramid has become larger (more people), and also more interconnected than ever before thanks to the rapidly improving means of communication. That has led to a speedier diffusion of both information and material culture, including tools, products of many kinds, plants, and animals, all of which increasing in nonlinear ways.

Compared to human chemistry, the biochemistry of life is still superior for providing free energy and matter in the form of food. That is why agriculture has not yet been replaced by industrial ways of food production. Yet today's industrial farms often use a higher input of fossil free energy than their resulting output in terms of food. This situation may continue to exist as long as fossil energy remains relatively cheap, while its biospheric effects are not considered too deleterious.

In sum: through the process of industrialization humans began to capture increasingly-diverse sources of free energy, while they started to do all kinds of things with them. As a result, the human food-capturing pyramid grew rapidly in size while undergoing profound changes. The same can be said for the domesticated food-capturing pyramid, most notably perhaps the growth of large mono cultures, and, more recently, the introduction of genetically-manipulated species. At the same time, the undomesticated food pyramid kept shrinking while it became less diverse.

WHAT HAPPENED TO HUMAN POPULATION GROWTH IN INDUSTRIAL TIMES?

Compared to earlier periods, during industrial times the human world population did not only reach staggering numbers, but also its doubling rate rapidly rose. According to Massimo Livi-Bacci, in 1750 CE about 771 million people would have been living on this planet, with a doubling rate of once per every seventy-two generations. In 1950 CE, two years before I was born, there were 2.5 billion people alive, while the doubling rate had increased to once per every eight generations. And during the period between 1950 and 2022 CE when this text was completed, the world population grew to about 7.96 billion people. This increase took place within only five generations of fifteen years each.

This means that during my lifetime, the doubling rate has speeded up to once per every three of such generations. Those are the highest ever human population growth rates during its entire history, while at the same time, more people have been alive than ever before. Those numbers show how quickly, in nonlinear ways, both the total population size and the doubling rates have risen since the beginning of the industrial revolution. A considerable portion of the population growth as of the twentieth century was caused by the fact that average human life-span almost doubled, due to more affluence and better health care, both of which were also unprecedented in human history.

Yet in today's affluent urban societies, birth rates have been dropping sharply, sometimes below reproduction levels, while currently 50% or more people live in cities as part of what has become known as the rural-urban transition. In city life, children are expensive and thus less desired, while joint and individual pension plans have lessened the urbanites' dependence on their children for their old-age income. Furthermore, for the first time in history a great many people have gained access to efficient forms of birth control. And not least of all, also the generation length has increased. Today, many urban women begin to have children, if at all, around thirty years of age, if not older, which implies a doubling of the earlier generation length.

As mentioned in the previous chapter, it is unclear to me, if, or to what extent, the effects of the Fifth Law may have counteracted those of the Fourth Law as a whole during this period. This remains an open question.

In sum: as a result of industrialization, the human numbers rapidly increased, while in more affluent areas this growth is now tapering off, if not being reversed. Yet in poorer mostly rural areas, birth rates often remain high. This leads to an increasing migration from those areas to the wealthier cities.

BIOSPHERIC EFFECTS OF THE INDUSTRIAL REVOLUTION

What were the most important novel biospheric effects of human action as a result of the industrial revolution? A great many studies exist that describe such global ecological influences. In addition to Vernadsky's *The Biosphere* and James Lovelock's books on Gaia, pioneering books include, in chronological order, Clive Ponting's *A Green History of the World* (1992); I.G. Simmons' *Changing the Face of the Earth* (1994); Neil Roberts' *The Holocene* (1998); John McNeill's *Something New Under the Sun: An Environmental History of the Twentieth Century World* (2000); Andrew Goudie's *The Human Impact on the Natural Environment* (2004); and Vaclav Smil's many detailed and insightful books about energy use in human history and their ecological consequences, including *The Earth's Biosphere* (2002). There are many more such books, including the ones about the biosphere's history mentioned at the beginning of Chapter 4.

Yet none of those studies employs the theoretical scheme advocated in this account. Seen from that perspective, during the industrial age the rapidly-increasing human capture of free energy joined hands with their capacity of biological nonlinear growth, and also with their unique cultural capacity of manipulating the rest of nature. All those aspects have a nonlinear growth potential. Yet by exploiting fossil fuels, humans also began to decrease their availability at an accelerating rate. In that respect, the biosphere moved a little more toward thermodynamical equilibrium than it otherwise would have been.

The industrial efforts introduced many novel biospheric effects. Within the context of this account, it is impossible to outline all of that in any detail. Even a reasonably complete summary appears out of reach. As a result, only a first attempt at writing such a summary is offered here.

In general, the desire for inanimate free energy led to the growing mining of such resources, in doing so moving the atoms and molecules involved that had been concentrated by natural processes, mostly carbon and hydrogen, to places where they were concentrated by humans and subsequently burned. All of that necessitated many types of artificial complexity. To be sure, such mining had already occurred previously, and those earlier human efforts should not be underestimated. But as of the industrial revolution, this began to occur worldwide on a much larger scale and with an increasing rate, in doing so producing growing biospheric effects.

Not only coal, oil, and gas were mined, but also all the other materials that were needed to make whatever humans wanted, mostly metals such as iron, copper, and aluminum, but also many other inorganic substances such as rocks, chalk, sand, salt, etc. The list is long, and began to include virtually all the known chemical elements, in doing so moving them to places where they otherwise would not have been.

Burning inanimate fuels led to increasing changes in the composition of the atmosphere, locally, regionally, and worldwide. Because coal and oil reserves contained nitrogen and sulfur, also those substances were burned, and the resulting acidic gases were released into the air, leading to growing effects around the globe. Also the simultaneous release of large amounts of dust into the atmosphere has had considerable biospheric effects. Furthermore, ever since people began to use the free energy stored inside large atoms such as uranium, many radioactive elements resulting from their use in nuclear reactors and bombs have spread within the biosphere.

But most importantly perhaps, the growing carbon dioxide content of the air as a result of human action began to influence the worldwide climate, because this higher concentration makes it more difficult for low-frequency radiation to disappear into the cosmos. This enhanced greenhouse effect warms up the lower portions of the atmosphere while it cools the higher-situated stratosphere, in doing so making it shrink. Furthermore, because of the melting ice caps in the polar zones and the resulting changes in global mass distribution, even the angle of Earth' axis would have been shifting as a result of human action. Before industrial times, the transition to agriculture and the concomitant lesser capture of carbon dioxide while releasing part of it into the air by burning the land, may also have had such effects, at least to some extent. But the industrial burning of the large amounts of coal and oil has been an altogether different affair, with altogether different effects.

Why would the increase of carbon dioxide in the air matter so much, while the simultaneous decrease of free oxygen is hardly ever mentioned? The reason for this is, that there is very little carbon dioxide in the air, because green life has been working so hard to suck it up for making its own bodies. Over the past 800,000 years, the carbon dioxide levels in the air would have fluctuated between 200 and 300 parts per million, so, between 0.02% and 0.03% of the atmosphere's composition.[1] Seen within this context, the increase of today's carbon dioxide concentration to 0.04% as a result of burning fossil fuels represents a substantial rise. By contrast, there is about 21% of oxygen in the air. So, a decrease of a fraction of a percent of it as a result of burning carbon and hydrogen hardly makes a difference.

All of that, including the rapidly increasing and diversifying forms of human construction and transportation of many kinds, is having enormous influences within the biosphere, many of which may still mostly be unknown. Not all those developments were totally new. For instance, people had started building and transporting things considerably earlier in time. But the scale, diversity, and speed of the industrial biospheric effects were unprecedented.

---

[1] https://www.climate.gov/news-features/understanding-climate/climate-change-atmospheric-carbon-dioxide (accessed May 3, 2021).

The rapidly-improving knowledge of chemistry allowed the creation and manipulation of an ever-growing variety of materials, including many products that had not existed before. As a result, such materials had not yet been part of any existing complex cif loops. Today, this includes as many as 350,000 different types of artificial chemical compounds, often with unknown effects on human, animal and plant health (Personn et al. 2022). Also those ways of manipulating the natural world had a longer history. But with the rise of the chemical industry, they acquired hitherto unknown dimensions through the large-scale production of substances such as plastics, fertilizers, pesticides, paints, glues, etc., while others chemicals such as, for instance, medicines and birth control pills, were made on a much smaller scale. Yet also some of those, at first sight, minor substances could still have considerable biospheric effects.

The production of materials that had not yet existed before may have a great many unpredictable effects, especially if they are damaging to other life-forms, and even more so when used in large quantities, such as plastics and pesticides. All of that began to influence the existing biospheric complex cif loops while creating new ones. In earlier times, far fewer novel substances had been made by humanity. Those earlier products could, therefore, more easily be relooped. Many industrial products, by contrast, began to consist of novel chemical compounds that are not easily degraded and relooped within the biosphere. As a result, they tend to accumulate and spread globally.

The industrialization of agriculture has been achieved by using machines powered by inanimate free energy; by using fertilizers, mostly nitrogen chemically extracted from the air, but also phosphorus and potassium, which are mined wherever they occur in sufficiently large quantities; by creating artificial environments such as greenhouses, heated by inanimate free energy; and by many types of powered irrigation schemes. All of that is aimed at maximizing the capture of solar energy by plants and convert it into forms of edible free energy and matter.

This has led to a situation in which today one single plant species, most notably perhaps maize for animal fodder, can jointly capture larger amounts of solar free energy than the undomesticated food-capturing pyramid would have been able to achieve in such an area without inputs such as machines, fertilizers and pesticides. And by introducing industrial agriculture, the soil content was changed, as was mentioned earlier, most notably perhaps affecting its microorganisms, but also influencing larger organisms such as Darwin's earthworms. All of that increasingly changed biospheric areas and the associated complex cif loops.

What happened in the oceans? The increased pressure on fishing incisively changed the aquatic food-capturing pyramids, leading to the depletion of many species as well as to efforts to domesticate the production of fishes in ponds, shallow bays, and the like. At the same time, the seven seas had become dumping grounds for products and refuse, including everything that ran off the land or descended into it from the air.

Furthermore, as a result of the growing use of electricity the biosphere began to light up at night, resulting in further mostly unknown effects. Earlier in time, humans had also produced forms of lighting at night, by burning wood, animal fat, oil, and gas. Also fireflies would have produced some illuminating effects. But all of that was minimal compared to how the current use of electricity is turning night almost into day in those areas where humans are concentrated in large numbers. This was the first time in the biosphere's history that such changes began to occur at such a large scale.

In addition, increasingly, societies are seeking to modify the weather, or its effects, by seeding clouds with silver iodide to stimulate rainfall or by bombing frozen Siberian rivers to break up the ice. No other species had done such things yet in the biosphere's history.

And perhaps not least of all, the industrial way of life has also produced many more sounds, ranging from the noises made by a large variety of machines to the sounds conveyed by loudspeakers. At the same time, many animal sounds diminished, because the species that made them declined in numbers. As a result, the biosphere's sound content rapidly changed. This made it more difficult, for instance, for whales to engage in long-distance communication by using sounds, because the noises made by ships cause too much disturbance in the oceans. Those noises may also disorient other sea animals, all with further biospheric consequences.

And all of that has been happening within only the most recent 0.000006% of the biosphere's history.

INTERMEZZO: PERSONALLY EXPERIENCING THE EMERGENCE OF THE INFORMATIZATION AGE

While performing my cultural-anthropological fieldwork in Peru in the 1980s and early 1990s, I stayed in contact with my Dutch family, friends, and academic supervisor mostly through sending letters and postcards. In those days, email and the internet did not yet exist, at least not in forms that were accessible to me. A period of four days was the postal record for a letter to arrive at Cusco's Central Post Office after having been posted in the Netherlands. Usually it took a little longer, often up to a week.

This meant that usually two weeks would pass between sending a letter and receiving an answer. That was considered fast and modern at that time. During the Spanish colonial period, such exchanges might have taken as much as two years. The increased speed of information exchange through letters had become possible thanks to the industrialization process, most notably thanks to daily commercial jet flights between Amsterdam and the Peruvian capital Lima as well as three to four daily flights between Lima and Cusco. Those internal flights to Cusco mostly transported tourists from Europe and the United States who were eager to visit the ancient Inca citadel of Machu Picchu.

Making a phone call from Peru to the Netherlands in the 1980s and early 1990s involved going to the central telephone office in Lima at Plaza San Martín; buying three minutes of phone call for ten dollars (a poor Peruvian laborer's weekly wage) at the box office; sitting in the waiting area together with many other people until the female operator would suddenly call through a creaky loudspeaker: ¡cabina once, para Holanda! I would hurry to that particular telephone booth, take up the horn, listen to the operator who first verified whether I was indeed the correct person, and then tried to make the connection. After some ringing, clicking, and other noises, I would suddenly hear the voices of my father and mother quite clearly, and we could talk for three minutes. What a miracle, what a luxury!

In those days, I kept informed of the world news mostly thanks to my little short-wave radio, which, if the atmospheric conditions were sufficiently good after nightfall, could receive both the BBC World Service and Radio Netherlands Worldwide Service. That is how I learnt for the first time, for instance, about the Chernobyl nuclear disaster in April of 1986. But other than that, I was basically living in Peru without any further contacts with Holland, while the opposite was the case after having returned to the Netherlands. No Peruvian news was available there on a daily basis, and no other ways of staying into contact either except by writing letters. For finding information about that country, I needed to go to the library of the University of Amsterdam's Latin America Institute.

Yet change was in the air. After returning from Peru to the Netherlands in May of 1986, I found that my Dutch acquaintances were excitedly familiarizing themselves with desktop computing. In Peru, I had just spent most of my time living with my Andean family in circumstances that, in many ways, resembled those of medieval rural Western Europe. Furthermore, I had studied partially decayed handwritten church documents, some of which dated back hundreds of years, which I had needed to copy by hand because digital photography did not yet exist. In those days, one would see the photographic results only after the negative films had been developed and printed. Because my access to those precious documents was sparse and unpredictable, I did not want to run the risk of photographic failure, and thus potentially losing any hard-won information. That is why I copied them by hand.

Yet back in Amsterdam in the fall of 1986, I suddenly found myself trying to master this brand-new computer technology: a desktop machine with a black and white screen, a keyboard, and two large floppy disk drives (no hard drives yet), that ran MS-DOS as the operating program as well as a word processor program. Before that time, I had written all my documents on typewriters. This was my first hands-on experience with the process of computer informatization as part of our daily lives. Yet when I returned to Peru in 1996 and 1997, email and the internet had arrived. From my rather privileged position as an honored guest at Cusco University (UNSAAC), I could send emails to Holland and might expect an answer the next time I went to that office, perhaps within a day. What a miracle!

Contrast that with today's situation. Using my rather average laptop computer, extraordinary amounts of information can now be accessed almost instantaneously. This includes a great many Peruvian newspapers and websites; several 'live' Peruvian TV and radio channels; including listening 'live' to the popular Cusco radio station Radio Salkantay, to which I had often listened in Peru. Furthermore, today a wealth of information about the village of Zurite is available on the internet, including many photos and videos. Before I went there for the first time in 1985, it had been impossible to find any information about that particular village in Amsterdam.

Through email, chats and video calls we can now exchange information with almost anyone in the world as long as that person has access to such technology. Many people now own smartphones with built-in digital cameras. This has led to a virtually endless stream of images and videos on the internet. And compared to earlier times, all of that is extraordinarily cheap. All these developments are part and parcel of the huge technological and social changes caused by the process of informatization.

The Internet cabin of the Universidad Nacional San Antonio Abad del Cusco on Avenida El Sol in Cusco, July 1997. (Photo by the author)

INFORMATIZATION: USING INANIMATE ENERGY FOR REPLACING HUMAN BRAINS BY COMPUTERS

Much like how steam engines began to replace human and animal muscle power as well as wind and water power, the introduction of computers in the 1950s and early 1960s heralded a clear acceleration of replacing human brains by such devices. And after digital computers became interconnected worldwide in the 1990s, the process of informatization rapidly began spreading around the world. Again, no other species in the biosphere's history had achieved any of that before. In our account, now we are dealing with the most recent 0.0000016% of the biosphere's history. Also the process of informatization has exhibited nonlinear, accelerating, characteristics as long as the social and ecological circumstances permit it. Within this context, it is important to note that information and culture have rarely been goals by themselves, if ever. They have been means for improving the survival chances of the people involved.

For observing all of that from a distance, there is no longer a need for imaginary visitors from space, because humans became such observers themselves, with their own eyes; through photos and movies; as well as through images captured by artificial satellites. Seeing Earth at a distance led to large changes in human perceptions, including pondering the questions of how our planet's biosphere works as well as what its history may have looked like. This led to a great many studies, including the present one.

The externalization of information storage from brains to artificial information carriers was not an entirely new process. A good case can be made that already paleolithic cave paintings represented the externalization of brain images, while the great many other images produced by humanity since that time can also be interpreted in such ways. One could even argue that such externalization processes had started millions of years earlier by making the first stone tools. The inventions of more elaborate forms of writing, printing, the telegraph, radio, telephones, television, most notably recording sounds and images with the aid of them or in other ways, added whole novel dimensions to that process. But in none of those cases the information was processed outside human brains. Calculators such as the Chinese abacus, astrolabes, and, more recently, slide rules, may offer exceptions.

The replacing of brains by programmable computers that processed information represented, therefore, an entirely novel survival strategy. Inevitably, this led to the adaptive radiations which we are currently experiencing, including the emergence of many kinds of computer-related businesses, while their successful owners are now among the wealthiest people within the biosphere. From a general point of view, computer hardware represents further refined forms of tool and fire use, while the growing numbers of interconnected people using the resulting efficient means of communication have produced another impressive acceleration in the diffusion of cultural information and material culture.

Another major effect has been the intensification of many already-existing industrial processes, including the replacement of humans by computer-steered processes. Within a short period of time, computers became able to perform a wide range of tasks at ever faster rates, and often with a greater precision than human brains could achieve. Early practical applications involved performing more complex calculations, including launching rockets and steering spacecraft, as well as communicating with them. This was followed by steering and controlling almost everything else as long as computers could do so better than human brains.

Today, computers also facilitate the almost instantaneous communication between people all around the globe as long as they are connected to the internet. This makes possible the diffusion of knowledge as well as other forms of human interaction at unprecedented speeds and scales, including the social media and playing computer games. As a result, many people have become less connected to other people through direct human contacts. The large production of data and the gathering of them for increasingly diverse purposes are other prominent aspects of our times. And so the list goes on.

The process of informatization has led to a considerable restructuring of the human food-capturing pyramid, including the emergence of novel human survival strategies and the decline, if not outright disappearance, of earlier ones. As an example, one may consider what happened to all the secretaries who up until the 1980s produced reports on typewriters, or how the newspaper industry has fared.

Yet at the same time, since the 1970s many aspects of daily life in affluent societies have not changed a great deal. For instance, in those years my parents' family owned a refrigerator, a car, a dishwasher, two telephones, several radios as well as a television set, while also jet travel was becoming more accessible. But we did not yet have computers or the internet, and not even small electronic calculators. The first exam that I needed to pass in 1970 while starting to study chemistry was a slide rule test. This was required for being able to performing basic calculations. That was not exceptional. At that time, even in Mission Control in Houston, Texas, Apollo engineers were still using slide rules for such purposes.

The intensification of the informatization process is closely related to the growing numbers of people as well as to their increasing wealth. When this process began to take off in the 1960s, there were about 3.5 billion people living on our planet, while in 2022 CE we are close to eight billion of us. As a result, today there are more, and more affluent, people living on this globe than ever before. And they also capture more free energy and matter than ever before, with the aid of which we make more stuff, and do more things, than ever before. To be sure, large numbers of poor people have remained, for whom capturing enough free energy and matter to survive the struggle for life has remained a daily experience.

BIOSPHERIC EFFECTS OF INFORMATIZATION

The biospheric effects of the informatization process can be summarized by saying that all the earlier-existing trends were intensified as a result of growing numbers of people who became globally interconnected, while more of them began to live in cities than ever before. The emergence of computers, smartphones, tablets, global networks, and data centers, as well as all the people associated with doing such things, and, of course, all the required infrastructure, necessitated the capture of more free energy and matter, which stepped up the effects of the Fifth Law. Those more-refined uses of electricity began to add growing amounts of electromagnetic radiation to the biosphere, for instance emanating from wires and transformers, but also increasingly in the form of transmissions that contained information such as radio, television, and telephone signals. This also influenced all the complex cif loops associated with those changes, while inevitably, this resulted in growing entropy as well, including increasing concentrations of carbon dioxide and other exhaust gases in the air, which influenced the associated complex cif loops. But because many earlier activities began to diminish, if not completely disappear, some decline of portions of the Fifth Law's effects also occurred.

The informatization process favors urban survival strategies, because such people have greater access to global networks and, in consequence, to all the opportunities that this offers. This has led to further rural decline, including the associated harvesting of free energy through agriculture, especially in mountainous areas, all with effects on the related complex cif loops.

During this period, there has been a growing awareness of all those issues, including attempts to do something about them. This includes recycling schemes of glass, plastic, paper, waste water, etc., as well as attempts to control the production of gasses that are considered harmful, such as CFC gases that affect the ozone layer and carbon dioxide as a major cause of human-induced climate change; and also the release of other gases including nitrogen and sulfur oxides which produce all kinds of biospheric effects.

Such recycling schemes are a novel aspect of the biosphere. No species had ever done so before. Because they all require free energy and matter to function, this has led to a further need for capturing free energy. But notwithstanding such noble efforts, today large amounts of human-made materials are still spreading through the biosphere. The current issues of micro plastics, fine dust, and radioactive waste offer such examples. Again, the list is long, and it is far beyond the possibilities of this account to provide an overview of all of that.

In sum: never before in the biosphere's history one single species has changed it to such an extent and within such a short period of time.

WHAT ABOUT SPACEFLIGHT AND THE SHIFTING BOUNDARY OF THE BIOSPHERE?

What about human efforts to leave the biosphere and go to places where no other Earthbound species had gone before? First of all, human spaceflight would not have been possible without computers and electronic wireless long-distance communication. The first crewed spaceflights may have involved only a few computers. But ever since those developments took off in the 1960s, computers have played a most important role in spaceflight, to the extent that the Apollo Project not only used increasingly-sophisticated computers and electronic communications, but that it also greatly stimulated their further development. This produced a remarkable lead by the United States in computer technology and the associated businesses during the decades that followed.

As part of that, humans began to move out of the confines of the biosphere, including sending men to the Moon and unmanned spacecraft to all the planets and beyond. Furthermore, during the past two decades the crewed International Space Station has been orbiting our biosphere. It was preceded by several earlier space outposts: the US Skylab and the Soviet Salyut and Mir space stations. Within this context, also the Chinese Tiangong space stations deserve to be mentioned, while new Russian and US space stations are now in the planning stage.

Does all of that mean that the biosphere has now become larger as a result of human action? If we use as a criterion the area within which biospheric complex cif loops are operating, the answer is: yes, because all of those space-related activities are influencing some of those loops back on Earth. They do so through all the human organizations that are needed to build, launch, and maintain them; through the rockets, and through their fuels that power them into the sky; by those portions of the rockets that fall back on Earth, crashing or soft landing; by a growing satellite belt with similar feedback effects through the electronic information gathered, sent back or relayed by them; by the transfer of people and their needs up and down the various space stations; by people landing on the Moon and leaving instruments there, of which some laser reflectors are still functioning; by landing spacecraft on Mars, Venus, and Saturn's moon Titan; by putting spacecraft in orbit around the Moon as well as around most other planets, or through passing them by, while some of those spacecraft are now moving out of the Solar System. As long as humans keep communicating with those crews and spacecraft on a regular basis, including all the required material needs and effects, they will have biospheric effects on Earth in the form of complex cif loops. As a result, those spacecraft and crewed missions should be seen as small outposts of the expanding biosphere of our home planet. And as soon as those spacecraft stop relaying information back to Earth, they will become biospheric fossils.

All of this leads to the inevitable conclusion that the age of informatization has enabled our species to extend Earth's biosphere considerably beyond its earlier boundaries, at least temporarily, in ways that no other species had done before. Surely, we live on an extraordinary planet, within an extraordinary biosphere, in extraordinary times.

What about a short summary of today's biosphere as well as of a projection of our species' future within it? That is the subject of the next and final chapter.

BIBLIOGRAPHY

Allen, Timothy F. H., Tainter, Joseph A. & Hoekstra, Thomas W. 2003. *Supply-Side Sustainability*. New York, Columbia University Press.

Alvarez, Walter. 2016. *A Most Improbable Journey: A Big History of Our Planet and Ourselves*. New York, W.W. Norton & Company.

American Geophysical Union. 2020. 'Ancient shell shows days were half-hour shorter 70 million years ago: Beer stein-shaped distant relative of modern clams captured snapshots of hot days in the late Cretaceous.' *ScienceDaily*. www.sciencedaily.com/releases/2020/03/200309135410.htm (accessed March 28, 2020).

Bellwood, Peter. 2005. *The Origins of Agricultural Societies*. Oxford, Blackwell.

Burroughs, William J. 2006. *Climate Change in Prehistory: The End of the Reign of Chaos*. Cambridge, Cambridge University Press (2005).

Carneiro, Robert L. 1970. 'A theory of the origin of the state.' *Science* 169, 3947 (733–738).

Carneiro, Robert L. 2012. 'The circumscription theory: A clarification, amplification, and reformulation.' *Social Evolution & History* 11, 2 (5–30). Moscow, 'Uchitel' Publishing House.

Chaisson, Eric J. 2001. *Cosmic Evolution: The Rise of Complexity in Nature*. Cambridge, MA, Harvard University Press.

Cochran, Gregory & Harpending, Henry. 2009. *The 10,000 Year Explosion: How Civilization Accelerated Human Evolution*. New York, Basic Books.

Cohen, Marc Nathan. 1977. *The Food Crisis in Prehistory: Overpopulation and the Origins of Agriculture*. New Haven & London, Yale University Press.

Cook, Earl. 1971. 'The flow of energy in an industrial society.' *Scientific American* 225, 3, September (134–147).

Cook, Earl. 1976. *Man, Energy, Society*. San Francisco, CA, W. H. Freeman & Co.

Crosby, Alfred W. 1972. *The Columbian Exchange: Biological and Cultural Consequences of 1492*. Westport, CN, Greenwood Press.

Crosby, Alfred W. 1993. *Ecological Imperialism: The Biological Expansion of Europe, 900–1900*. Cambridge, Cambridge University Press (1986).

Crosby, Alfred W. 2006. *Children of the Sun: A History of Humanity's Unappeasable Appetite for Energy*. New York, W.W. Norton & Company.

Darwin, Charles. 1859. *On the Origin of Species by Means of Natural Selection, or the Preservation of Favoured Races in the Struggle for Life*. London, John Murray.

Darwin, Charles. 1881. *The Formation of Vegetable Mould, through the Action of Worms, with Observations on Their Habits*. London, John Murray.

Darwin, Charles, Fitz-Roy, Robert, & King, Phillip Parker. 2015a. *Narrative of the Surveying Voyages of His Majesty's Ships Adventure and Beagle between the Years 1826 and 1836*. Volume 2, Proceedings of the Second Expedition, 1831–1836. Cambridge, Cambridge University Press (1839).

Darwin, Charles, Fitz-Roy, Robert, & King, Phillip Parker. 2015b. *Narrative of the Surveying Voyages of His Majesty's Ships Adventure and Beagle Between the Years 1826 and 1836*. Volume 3. Journal and Remarks 1832–1836. Cambridge, Cambridge University Press (1839).

Diamond, Jared. 1997. *Guns, Germs and Steel: The Fates of Human Societies*. London, Jonathan Cape.

Droste, Dietrich. 2010. *Energiemangel als Antrieb der Menschheitsgeschichte: Eine energetische Gesellschafts- und Geschichttheorie*. München, Martin Meidenbauer Verlagsbuchhandlung.

Eisenstein, Elizabeth L. 1993. *The Printing Revolution in Early Modern Europe*. Cambridge, Cambridge University Press (1983).

Elhacham, Emily, Ben-Uri, Liad, Grozovski, Jonathan, Bar-On, Yinon M., & Milo, Ron. 2020. 'Global human-made mass exceeds all living biomass.' *Nature* 588, 17 December (442–444) https://www.nature.com/articles/s41586-020-3010-5.

Elias, Norbert. 1978. *What is Sociology?* London, Hutchinson.

Fieldhouse, David Kenneth. 1973. *Economics and Empire, 1830–1914*. London, Weidenfeld & Nicolson.

Flynn, Dennis O. & Giráldez, Arturo. 1995. "Born with a 'silver spoon": The origin of world trade in 1571.' *Journal of World History* 6, 2 (201–222).

Flynn, Dennis O. & Giráldez, Arturo. 2002. 'Cycles of silver: Global economic unity through the mid-eighteenth century.' *Journal of World History* 13, 2 (391–428).

Flynn, Dennis O. & Giráldez, Arturo. 2008. 'Born again: Globalization's sixteenth century origins (Asian/global versus European dynamics).' *Pacific Economic Review* 13, 3 (359–387).

García Martínez, Adolfo. 2017. *Alabanza de aldea*. Oviedo, Krk ediciones.

Goudie, Andrew. 2004. *The Human Impact on the Natural Environment*, Fifth Edition. Malden, MA & Oxford, UK, Blackwell Publishing (2000).

Goudsblom, Johan. 1992. *Fire and Civilization*. London, Allen Lane.

Hanson, Robin. 2000. Long-term growth as a sequence of exponential modes. Unpublished paper: http://mason.gmu.edu/~rhanson/longgrow.pdf.

Harris, David R. (ed.). 1996. *The Origins and Spread of Agriculture and Pastoralism in Eurasia*. London, UCL Press.

Harris, Marvin. 1975. *Culture, People, Nature: An Introduction to General Anthropology*. New York, Harper & Row.

Heiser, Charles B., Jr. 1990. *Seed to Civilization: The Story of Food*. Cambridge, MA, Harvard University Press.

Hobsbawm, E. J. 1968. *Industry and Empire*. Harmondsworth, Penguin.

Humboldt, Alexander von. 1995. *Personal Narrative of a Journey to the Equinoctial Regions of the New Continent (Abridged and Translated with an Introduction by Jason Wilson and a Historical Introduction by Malcolm Nicolson)*. London, Penguin Books (1814–1825).

Humboldt, Alexander von. 1997. *Cosmos, vol. 1, Foundations of Natural History*. Baltimore, Johns Hopkins University Press (1858).

Jiménez de la Espada, don Marcos. 1965. *Relaciones Geográficas de Indias, II*. Madrid, Biblioteca de Autores Españoles, Ediciones Atlas (199–205).

Johnson, Steven. 2021. *Extra Life: A Short History of Living Longer*. New York, Riverhead Books.

Kortlandt, Adriaan. 1972. *New Perspectives on Ape and Human Evolution*. Amsterdam, Stichting voor Psychobiologie.

Kortlandt, Adriaan. 1980. 'How might early hominids have defended themselves against large predators and food competitors?' *Journal of Human Evolution* 9, 2 (79–112).

Kortlandt, Adriaan &. Coppens, Yves. 1994. 'Rift over origins.' *Scientific American* 271, 4 (5).

Landes, David Saul. 1969. *The Unbound Prometheus: Technological Change and Industrial Development in Western Europe from 1730 to the Present*. Cambridge, Cambridge University Press.

Landes, David Saul. 1998. *The Wealth and Poverty of Nations: Why Some Are So Rich and Some So Poor*. New York, W.W. Norton & Company.

Leakey, Richard E. & Lewin, Roger. 1995. *The Sixth Extinction: Patterns of Life and the Future of Humankind*. New York, Doubleday.

Livi-Bacci, Massimo. 1999. *A Concise History of World Population*, Second Edition. Cambridge, MA & Oxford, Blackwell (1997).

Lovelock, James E. 1987. *Gaia: A New Look at Life on Earth*. Oxford & New York, Oxford University Press (1979).

Lovelock, James E. 1995. *The Ages of Gaia: A Biography of Our Living Earth.* New York, W.W. Norton & Company (1988).

Lovelock, James E. 2000. *Gaia: The Practical Science of Planetary Medicine.* Oxford & New York, Oxford University Press (1991).

Malthus, Thomas. 1798. *An Essay on the Principle of Population, as it Affects the Future Improvement of Society, with Remarks on the Speculations of Mr. Godwin, M. Condorcet, and Other Writers.* London, Printed for J. Johnson, in St. Paul's Church-Yard. 1998, Electronic Scholarly Publishing Project: http://www.esp.org/books/malthus/population/malthus.pdf.

Markley, Jonathan. 2009. 'A child said, "What is the grass?": Reflections on the big history of the Poaceae.' *World History Connected* 6, 3. http://worldhistoryconnected.press.illinois.edu/6.3/markley.html.

McNeill, John R. 2000. *Something New Under the Sun: An Environmental History of the Twentieth Century World.* London, Penguin Books.

McNeill, John R. 2010. *Mosquito Empires: Ecology and War in the Greater Carribean, 1620–1914.* Cambridge, Cambridge University Press.

McNeill, John R. 2020. *The Webs of Humankind: A World History (Vol. 1).* New York, W.W. Norton & Company.

McNeill, William H. 1963. *The Rise of the West: A History of the Human Community.* Chicago & London, University of Chicago Press.

McNeill, William H. 1974. *The Shape of European History.* New York, London, & Toronto, Oxford University Press.

McNeill, William H. 1976. *Plagues and Peoples.* Garden City, N.Y, Anchor Press/Doubleday.

McNeill, William H. 1984. *The Pursuit of Power: Technology, Armed Force and Society since AD 1000.* Chicago, IL, University of Chicago Press (1982).

McNeill, John R. & Engelke, Peter. 2016. *The Great Acceleration: An Environmental History of the Anthropocene since 1945.* Cambridge, MA, The Belknap Press.

McNeill, John R. & McNeill, William H. 2003. *The Human Web: A Bird's Eye View of World History.* New York, W.W. Norton & Company.

Moffa-Sánchez, Paola & Hall, Ian R. 2017. 'North Atlantic variability and its links to European climate over the last 3000 years.' *Nature Communications* 8 (1726). DOI: 10.1038/s41467-017-01884-8.

Mottl, Ondřej, et al. 2021. 'Global acceleration in rates of vegetation change over the past 18,000 years.' *Science* 372 (860–864).

Mumford, Lewis. 1961. *The City in History: Its Origins, Its Transformations, and Its Prospects.* New York, Harcourt, Brace, and World Inc.

Nazaretyan, Akop P. 2004. *Civilization Crises within the Context of Big (Universal) History: Self-Organization, Psychology, and Futorology (in Russian).* Moscow, Mir-Publishers.

Nazaretyan, Akop P. 2010. *Evolution of Non-Violence: Studies in Big History, Self Organization and Historical Psychology.* Saarbrucken, Germany, LAP Lambert Academic Publishing.

Niele, Frank. 2005. *Energy: Engine of Evolution.* Amsterdam, Elsevier/Shell Global Solutions.

Odum, Howard T. 1971. *Environment, Power and Society.* New York, John Wiley & Sons, Inc.

Odum, Howard T. & Odum, Elizabeth C. 1981. *Energy Basis for Man and Nature*, Second Edition. New York etc., McGraw-Hill Book Company.

Persson, Linn, et al. 2022. 'Outside the Safe Operating Space of the Planetary Boundary for Novel Entities.' *Environmental Science & Technology*, Article ASAP, DOI: 10.1021/acs.est.1c04158.

Pinker, Steven. 2011. *The Better Angels of Our Nature: Why Violence Has Declined.* New York, Viking Press.

Pollard, Sidney. 1992. *Peaceful Conquest: The Industrialization of Europe 1760–1970.* Oxford & New York, Oxford University Press (1981).

Ponting, Clive. 1992. *A Green History of the World.* Harmondsworth, Penguin Books.

Potts, Rick. 1996. *Humanity's Descent: The Consequences of Ecological Instability.* New York, William Morrow & Co.

Rathje, William & Murphy, Cullen. 1992. *Rubbish! The Archaeology of Garbage: What Our Garbage Tells us about Ourselves.* New York, HarperCollins Publishers.

Roberts, Neil. 1998. *The Holocene: An Environmental History*, Second Edition. Oxford, Blackwell.

Rodríguez González, Agustín Ramón. 2018. *La primera vuelta al mundo, 1519–1522.* Madrid, EDAF.

Sherratt, A. 1981. 'Plough and pastoralism: Aspects of the secondary products revolution.' In: Hodder, Ian, Isaac, Glynn, & Hammond, Norman (eds.) *Pattern of the Past. Studies in Honour of David Clarke.* Cambridge, Cambridge University Press (261–305).

Simmons, Ian Gordon. 1994. *Changing the Face of the Earth: Culture, Environment, History.* Oxford, Blackwell (1989).

Slicher van Bath, Bernhard Hendrik. 1963. *The Agrarian History of Western Europe, A.D. 500–1850.* London, Edward Arnold.

Smil, Vaclav. 1994. *Energy in World History.* Boulder, CO., Westview Press.

Smith, Bruce D. 1995. *The Emergence of Agriculture.* New York, W. H. Freeman & Co., Scientific American Library.

Spier, Fred. 2015. *Big History and the Future of Humanity, Second Edition.* Oxford, Wiley-Blackwell.

Stearns, Peter N. 1993. *The Industrial Revolution in World History.* Boulder, CO, Westview Press.

Thomas, Julia Adeney, Williams, Mark, & Zalasiewicz, Jan. 2020. *The Anthropocene: A Multidisciplinary Approach.* Cambridge, UK, Polity Press.

Trivedi, Bijal. 2007. 'Toxic cocktail.' *New Scientist*, issue 2619, 1 September (44–47).

Tylor, Edward Burnett. 1871. *Primitive Culture: Researches into the Development of Mythology, Philosophy, Religion, Art, and Custom, Volume 1.* London, John Murray.

Vélez, Antonio. 2013. *Homo Sapiens: Psychology as Seen from Evolution.* Bogotá, Colombia, eLibros Editorial.

Vries, Bert de & Goudsblom, Johan. 2002. *Mappae Mundi: Humans and Their Habitats in a Long-Term Socio-Ecological Perspective, Myths, Maps and Models.* Amsterdam, Amsterdam University Press.

White, Leslie Alvin. 1943. 'Energy and the evolution of culture.' *American Anthropologist* 45 (335–356).

White, Leslie Alvin. 1959. *The Evolution of Culture: The Development of Civilization to the Fall of Rome.* New York, McGraw-Hill.

Williams, Mark, Zalasiewicz, Jan, Haff, P.K., Schwägerl, Christian, Barnosky, Anthony D., & Ellis, Erle. 2015. 'The Anthropocene biosphere' *The Anthropocene Review* 2, 3 (196–219).

Winter, Niels W. de, et al. 2020. 'Subdaily-Scale Chemical Variability in a Torreites Sanchezi Rudist Shell: Implications for Rudist Paleobiology and the Cretaceous Day-Night Cycle.' *Paleoceanography and Paleoclimatology*, 35 (2) doi: 10.1029/2019PA003723.

Wolf, Eric R. 1982. *Europe and the People without History.* Berkeley, University of California Press.

Zalasiewicz, Jan & Williams, Mark. 2012. *The Goldilocks Planet: The Four Billion Year Story of Earth's Climate.* Oxford, Oxford University Press.

8 What about Our Long-Term Survival Strategy within the Future Biosphere?

> In Papua New Guinea one would expect that tribal groups did not harvest the sweet potatoes all at once, so that, with a certain planning and labor, food would always be available. [However] I have witnessed several times that Tribal Communities harvested all the available food once it was Party Time. After that, those communities suffered from hunger, and borrowed or stole sweet potatoes from neighboring Tribes to plant them for the next harvest, three to four months later.
>
> Hanz Spier (my younger brother, personal email April 26, 2021, translated by the author from the Dutch, quoted with permission)

HOW CAN TODAY'S BIOSPHERIC SITUATION BE SUMMARIZED?

What is the state of today's biosphere, our large but limited common home within the extremely large and very inhospitable cosmos? And what could we do in trying to assure the long-term survival of our species within it? Those are the two leading questions that will be addressed in this final chapter.

Seen from a general point of view, humanity is currently facing issues that are similar to those of the Papua New Guinea tribal farmers mentioned above. However, our problems are worse, because today, we may be consuming most, if not all the available resources, and we may subsequently go hungry forever. After having plundered and spoiled our biosphere to the extent that it will no longer be able support a sizable human population in any well-being, there will not be any neighbors from whom such resources can be borrowed or stolen. Like the early agriculturists who needed to learn to think further ahead while adopting novel forms of self-control, for a successful common future in reasonable well-being all of humanity will need to do so, too (cf. Spier 2019).

How can our biospheric situation of today be described? Currently, the largest human numbers ever are inhabiting it, with recently also their highest demographic and economic growth rates ever, while we jointly have more material needs and ambitions than ever before. In this situation, we are joined by an estimated 26 billion chicken, 1 billion cattle, and 650 million pigs. Annually, we jointly consume ~2.2 billion metric tons of grain (including rice and maize, but not human-managed grasslands consumed by cattle) as well as 368 metric tonnes of potatoes.[1]

[1] These data were found in 2020 on www.statista.com/statistics.

DOI: 10.1201/9781003275350-8

Most of the industrial production is taking place in factories, which are often situated half a globe away from where their products are used. During the production process, half-products may also travel large distances until they end up within the finished products. All of that requires capturing large amounts of free energy; intensive mining for resources; moving all of that to the factories; making those things; and transporting them to the customers; while dealing with the inevitable refuse, the increasing entropy. As a result, today the total mass of human-made products exceeds the mass of all living beings combined (Elhacham et al. 2020). Other transportation needs, including tourism, require similar inputs.

In industrial farming, by preferring fast-growing crops and animals with large yields, humans are currently pushing the nonlinear growth capacity of domesticated life to its limits. This offers another example of the Fifth Law in action. And even though wild fires keep occurring, most fires within today's biosphere now appear to be caused by human action, planned and unintentional. At the same time, the undomesticated food-capturing pyramid keeps declining under human pressure, both on land and in the oceans.

As part of that, our species has parceled up most of the face of the Earth into areas controlled by human action. They are usually covered by plants or artificial materials such as buildings and roads, and by lots of cars as well as some trains, without which human intervention would not be able to reach those areas. Ships carrying free energy in the form of oil, coal, and liquefied gas ply the oceans, while other vessels transport a wide range of agricultural and industrial products, not to mention the waste generated by human action. High-voltage tension cables increasingly span the Earth transporting electricity to increasing numbers of people, but not to all. The electricity is generated in large plants; with the aid of dammed reservoirs; and, increasingly also though the capture of free energy present in sunlight and the winds, all in an effort to provide sufficient free energy to perform all those tasks.

As part of those developments, humans have been changing the land in unprecedented proportions. Already in 1995, for instance, it was reported that the pace of natural geological change in the U.S.A. was outdone by the rapid metamorphosis of the landscape through human enterprise. The US geologist Roger Hooke claimed that 'mining, road building, and construction of homes sift roughly 7.6 billion tonnes of soil a year in the US. By comparison, [US] rivers, the most important natural agents, transport only about 1 billion tonnes each year, if you ignore sediment dumped into them by human activity' (quoted in: *New Scientist* 4 March 1995, p. 11).

Today, water pipes transport this purified commodity to increasing numbers of people, but not to all, while other pipes take care of the discharges. Through irrigation and other forms of water control, humans are increasingly influencing the biospheric water loop. Current human water use would amount to 24,000 km^3 per year, which equals half of the effluent water mass of all rivers combined (Wang et al. 2018). And between 2002 and 2016, through human action water has systematically been moved from the land to the oceans. As a result, the amount of fresh water on land has decreased by, on average, 164 km^3 per year (Abbott et al. 2019).

Glass-fiber cables span the globe carrying ever-growing amounts of information. In the sky, planes interconnect all portions of the planet, while high above them, artificial satellites do the same, while looking down upon us and collecting various types of information for various purposes. And while from about 2.6 million years ago the biosphere's climate started oscillating as a result of the onset of the first ice ages, the biospheric effects of planetary gas emissions resulting from human action may be terminating this oscillation, at least temporarily, in doing so returning the biosphere to its former warmer conditions.

All those human actions, and a great many more, have been transporting atoms, molecules, and associations of molecules, old and new, to places where they would otherwise not have been. This has led to unprecedented changes within today's biosphere that are hard to chart, and even more difficult to understand, with any precision. And much of that began to happen between a mere 200 and 50 years ago. Yet this current world is almost taken for granted by many people in more affluent societies, especially by the younger generations, because our life-spans are shorter than most of those developments, while in many cases, our historical education has not kept pace with those biospheric changes.

As a result, our unappeasable appetite for free energy, summarized in this book as the Fifth Law of Thermodynamics, may be reaching its limits. Seen from a human perspective, its effects are increasingly spoiling the biosphere in many ways. Among them, the steadily-increasing greenhouse effect stands out as a result of the growing amounts of carbon dioxide released into the air. This situation is beginning to act as a break on harvesting carbon-based free energy, and, as a result, on the Fifth Law.

Interestingly, there appears to be a close correspondence between the increase in doubling rates of the human population during its history and the simultaneously-occurring rise of the estimated free energy capture per capita, as shown in the following table.

TABLE 8.1

Energy Use, Population Doubling Rates, and Population Growth

	Energy Use in Watt per Capita	Doubling Rate of Population Growth in Number of Generations (of fifteen years)	Population Size
Hunting and gathering (fire use excepted)	200	400	1–10 million (40,000 years ago)
Early agricultural societies	480	300	25–50 million (10,000 BP)
Advanced agricultural societies	1,040	100 → 30	500 million (1,500)
Industrial societies	3,080	70 → 8	800 million (1776)
Technological societies	9,200	3 (twentieth century)	8 billion (2022)

Sources: Energy consumption per capita in human history: Cook (1971, p. 136); doubling rates and population growth: Livi-Bacci (1999, p. 31). The terms used for characterizing human societies come from Cook (1971).

I am not aware of a publication that has already drawn attention to this remarkable situation. No other species in the biosphere's history appears to have done anything similar.

Even though the population doubling rates are currently sharply declining as a result of urbanization, the enormous current human numbers together with their material needs and desires will make sure that the Fifth Law will keep functioning in the foreseeable human future.

WHAT ABOUT THE DEFINITION OF THE ANTHROPOCENE?

As a result of all those developments, the period starting in 1800 CE is increasingly called the Anthropocene, following a suggestion in 2000 CE by the Nobel laureate Dutch atmospheric scientist Paul Crutzen (1933–2021). This would be the first geological period for which no clear marker can be placed within the rock record, because it takes longer than a few hundred years for rocks to form. Geologists call such reference points 'Global Boundary Stratotype Section and Points.'

Even though the lack of an official rock marker makes geologists reluctant to accept the Anthropocene as a geological epoch, a formal proposal for considering it as such was made in 2008 CE by the British geologist Jan Zalasiewicz and colleagues, because the current human influence on the biosphere is undeniable, and future generations would be able to drive a nail in sedimentary rocks pinpointing this period, if humanity survives long enough to do so.

All of this has led to discussions about what acceptable geological traces would be for signaling the beginning of the Anthropocene. Those signals may include radioactive traces resulting from nuclear explosions, or underground structures such as metro tunnels, which may survive for longer periods of time because they are less likely to erode underground than human constructions above ground (cf. Zalasiewicz 2008, Williams et al. 2015, Thomas et al. 2020).

Yet seen from the perspective of this book, the geological perspective for defining the Anthropocene view may be too narrow. As mentioned, in Chapter 4, processes often began earlier in time than the first surviving evidence at our disposal. And in the second place, would we perhaps need to further broaden our view and begin discussing the Anthropocene not only as a new geological period, but also as a novel biospheric period?

Thinking along those lines, we may want to broaden our definition of what such human leftover traces would be like, not only in rocks but perhaps also elsewhere. So, why would we only look down to what we can find in rocks, and not also up into the sky? Doing so may be unusual for geologists, because it does not make a great deal of sense for geological history.

Yet for the biosphere's history, that may be different. As of the late 1950s, in the sky above us an artificial belt has emerged that consists of increasing numbers of objects orbiting our planet, satellites as well as growing amounts of space debris. Much of that, especially in higher Earth orbit, may stay there for many millions of years, if not longer, so perhaps long after humans have gone extinct. Seen from such a future point of view, those artificial space objects or their leftover remains can be considered biospheric fossils that signaled the beginning of a once novel biospheric period.

However, by defining the Anthropocene in such a way, its duration right now would be only the most recent 0.0000016% of the biosphere's history which coincides with the period of informatization. Yet as explained in the previous chapter, the first emergence of specific human influences within the biosphere, so, in consequence, the beginning of the Anthropocene, may have occurred much earlier in time, perhaps as long ago as 1.5 million years, when our species first began using fire. In doing so, those early humans began to engage in activities that no other animals had done before. This initiated novel biospheric processes which, over the course of time, would profoundly alter the biosphere's composition.

Yet those early human activities did not leave any traces in rocks or elsewhere. As a result, geologists might not find such an early definition of the Anthropocene acceptable. But seen from a biospheric process-oriented approach, it appears more realistic to define the beginning of the Anthropocene as of that early period. Such a choice would be my current preference. And if this is not considered sufficiently convincing, one might consider the beginning of agriculture and animal husbandry, when human began to profoundly alter the biosphere, to be the start of the Anthropocene. Similar suggestions were put forward by the Dutch academics Bert de Vries and Johan Goudsblom in their book *Mappae Mundi* (2002). Thinking along those lines implies that we may need to redefine the biosphere's history in terms of biospheric periods, within which the current geological periods will, of course, loom large.

Whatever will be decided, there can be no doubt that humans have been influencing the biosphere in ways that are unlike those of any other species, over a much longer period of time, than the current definition of the Anthropocene's duration allows. Our knowledge of those processes is still insufficient. Yet our future will depend on understanding them better.

WHAT ABOUT FORECASTING THE FUTURE IN TERMS OF BIOSPHERIC TIME?

Could the future be forecast – the term 'forecast' was invented by captain Fitz-Roy – for longer periods of time, for instance the next 1,000 years, so, the next 0.000015% of the biosphere's history? Even if we could, planning for such a relatively short biospheric period appears impossibly long, seen from the current political perspectives, which are still mostly based on, at best, the duration of a human life-span, and, quite often, much shorter than that.

To clarify this conundrum, let us compare 1,000 years of biospheric planning with planning for my own personal life. I am now almost seventy years old. At my current age, contemplating the next 1,000 years of the biosphere's future would be equal to thinking about the next nine minutes of my life. That is not a long time, about as long as it took me to think of this comparison and perform the calculation. Yet this is about the longest period in time currently considered while trying to forecast our common human future on this planet. And if we compared my lifetime with the period that has elapsed since human began using fire, about 1.5 million years ago, thinking about the next 1,000 years would be equal to contemplating the next seventeen days of my life.

None of that is the kind of planning on which I would rely for my personal future. That would look pretty short-sighted, wouldn't it? Yet that is the type of planning that we are currently doing, or even much less, regarding our common future as a species. Clearly, if we want to achieve a more sustainable future in reasonable well-being for as many people as possible, we will need to learn to think in biospheric time and incorporate that into our common human survival strategy.

Seen from a long-term perspective, the history and future of human affairs within the biosphere may look much like the Eiffel Tower: a sudden steep climb to a top, quite likely to be followed by a similarly steep decline. If we want to live with a certain degree of sustainability within our biosphere – the only one that we have – it is our task, as well as the task of those who will come after us, to try to prevent this decline from happening, at least in such a steep way. It will be very difficult for humanity to reach a more sustainable future within this very dynamic biosphere. Yet doing so without causing biospheric effects that further damage our longer-term survival on this planet is the major challenge that humanity is facing today.

This challenge may be too difficult for our species, being what it is, including the socio-ecological conditions we find ourselves in today. If so, seen on the time-scale of the biosphere's history, human life on our planet will probably be very temporary. As a result, the Anthropocene would potentially become by far the shortest geological epoch in the biosphere's history. After humans have disappeared from the biosphere, the Fourth Law will reign supreme again until another species emerges that can revive the Fifth Law. Yet almost inevitably, major cosmic impacts will occur somewhere in the future that may severely damage life's existence on our planet. And after a few more billions of years, the end of the Sun's existence is expected to obliterate any still-existing life on our planet, and, in consequence, also Earth's biosphere. We cannot plan for dealing with such contingencies.

WHAT COULD WE DO? LONG-TERM VERSUS SHORT-TERM SURVIVAL STRATEGIES

As we have seen, in the biosphere's history humans are a very recent phenomenon. Previously, the continuously-evolving biospheric circumstances tended to marginalize, or even eliminate, all those species that could not sufficiently survive and adapt to the new circumstances. So, can we expect, in James Lovelock's terms, that Gaia will eliminate from the biosphere humans because of their damaging effects? Or would there be a process of natural selection that facilitates those humans to survive who can sufficiently adapt to the new circumstances? If so, how painful would that be? Could we avoid such pains?

In a general sense, the possibilities and limitations for future human action as I understand them were outlined in Chapter 8 of my book *Big History and the Future of Humanity, Second Edition* (2015). I see, therefore, no need to repeat those arguments here. But what may need some further elaboration is the time-span in which humans tend to think about the future, which may not be sufficient for our survival as a species in reasonable well-being.

There can be little or no doubt that for dealing successfully with all those future issues we will need to engage in long-term planning. Emphasizing such a need is not new at all. Yet rarely has it been turned into a systematic effort to be undertaken by all of humanity. There are exceptions, such as trying to deal with the ozone hole in the atmosphere as well as the current effects to mitigate climate change caused by human action. But if we want to achieve our longer-term survival in reasonable well-being, much more needs to be done. We will need to evolve long-term all-round survival strategies for human sustainability and implement them.

Doing so would be another novelty within the biosphere's history. The survival strategies of all the other species that have lived on this planet have, at most, involved the duration of their individual life-spans, including the next generations. That is what pepper plants do, as we have seen. They live during one single growth season and make the best of it, while preparing for the coming lean season. That is their life-span. Humans usually think, at best, in terms of the duration of our own life-span, including the next generation, but often much less than that, depending on the circumstances.

Doing so will, of course, remain important. Yet our successful survival as a species will require a survival strategy that is designed with a far longer time span in mind. Can we do that? What may favor doing so, and what may stand in the way? It is a hopeful sign that the human capacity of making overviews while paying attention to details is more advanced that anything achieved by other animals. This human capability may be crucial for making such plans and incorporate them into our survival strategies. To be successful, we would need to foster long-term approaches first of all in our education, but also within our political discourse and subsequent actions.

As part of that, long-term views of a hopefully more sustainable future will need to be enshrined in law. In that respect, there are some hopeful signs. In the German Constitution of 1994, Article 20a includes the phrase:[2]

> Mindful also of its responsibility toward future generations, the state shall protect the natural bases of life by legislation and, in accordance with law and justice, by executive and judicial action, all within the framework of the constitutional order.

The Bolivian constitution of 2009 states in Article 9:[3]

> Promover y garantizar el aprovechamiento responsable y planificado de los recursos naturales, e impulsar su industrialización, a través del desarrollo y del fortalecimiento de la base productiva en sus diferentes dimensiones y niveles, así como la conservación del medio ambiente, para el bienestar de las generaciones actuales y futuras.

Ecuador's constitution of 2015 states in its Preamble:[4]

> CELEBRANDO a la naturaleza, la Pacha Mama, de la que somos parte y que es vital para nuestra existencia,

and in Article 395:

> La Constitución reconoce los siguientes principios ambientales: 1. El Estado garantizará un modelo sustentable de desarrollo, ambientalmente equilibrado y respetuoso de la diversidad cultural, que conserve la biodiversidad y la capacidad de regeneración natural de los ecosistemas, y asegure la satisfacción de las necesidades de las generaciones presentes y futuras.

---

[2] The 1994 German constitution (in German): https://www.gesetze-im-internet.de/gg/BJNR000010949.html and in English: https://www.bundesregierung.de/breg-en/chancellor/basic-law-470510.

[3] The constitution of Bolivia (2009): http://www.oas.org/dil/esp/constitucion_bolivia.pdf.

[4] Constitution of Ecuador (2015): https://www.ambiente.gob.ec/wp-content/uploads/downloads/2018/09/Constitucion-de-la-Republica-del-Ecuador.pdf.

From my personal point of view, those are hopeful signs. There may be more states around the world who have enshrined such principles within their laws. All of that may herald the beginning of fostering long-term biospheric views, including turning such thoughts into action with a legal, institutional, basis. In terms of Norbert Elias, such a novel behavior would need to be turned into forms of self-control that become so ingrained that they appear 'natural.' As part of that, a worldwide collaboration at an unprecedented scale will be required.

Much more can be said about all these topics. But in sum: whether we like it or not, humans will have to learn to live with the current situation well summarized in the biblical account of creation mentioned in Chapter 6, in which we have acquired 'dominion over the fish of the sea, and over the fowl of the air, and over the cattle, and over all the earth.' By the time those words were formulated, perhaps as early as 4,000 years ago, that Middle Eastern area had already experienced 6,000 years, if not more, of agriculture and animal husbandry. So, for those people that situation had become the norm, and they realized it. It is about time that we realize this, too, but today in our own ways, based on the best available empirical knowledge and scholarly interpretations of our biosphere' history.

In conclusion, if we want to promote the longer-term sustainable survival of our species in reasonable well-being for as many people as possible, we will have to learn to manage all of that well, with modesty, wisdom, and respect for the biosphere as a whole. To be able to do that sufficiently well, we will need to understand the biosphere's history and present, while we would also need to learn to think ahead in terms of biospheric time. If the approach proposed in this book will help us to take a few steps along that road, I will be more than happy.

BIBLIOGRAPHY

Abbott, Benjamin W., et al. 2019. 'Human domination of the global water cycle absent from depictions and perceptions.' *Nature Geoscience* 12 (533–540). DOI: 10.1038/s41561-019-0374-y.

Cook, Earl. 1971. 'The flow of energy in an industrial society.' *Scientific American* 225, 3 (134–147).

Elhacham, Emily, Ben-Uri1, Liad, Grozovski, Jonathan, Bar-On, Yinon M. & Milo, Ron. 2020. 'Global human-made mass exceeds all living biomass.' *Nature* 588 (442–444). DOI: 10.1038/s41586-020-3010-5.

Livi-Bacci, Massimo. 1999. *A Concise History of World Population*, Second Edition. Cambridge, MA & Oxford, Blackwell (1997).

Lovelock, James E. 1987. *Gaia: A New Look at Life on Earth*. Oxford & New York, Oxford University Press (1979).

Lovelock, James. 1995. *The Ages of Gaia: A Biography of Our Living Earth*. New York, W. W. Norton & Company (1988).

Lovelock, James. 2000. *Gaia: The Practical Science of Planetary Medicine*. Oxford & New York, Oxford University Press (1991).

Lovelock, James E. 2006. *The Revenge of Gaia: Why the Earth Is Fighting Back and How We Can Still Save Humanity*. London, Allen Lane.

McNeill, John R. & Engelke, Peter. 2016. *The Great Acceleration: An Environmental History of the Anthropocene since 1945*. Cambridge, MA, The Belknap Press.

New Scientist. 1995. ‘People outdo the elements as shapers of the landscape.’ *New Scientist,* issue 1967, 4 March 1995, (11).

Spier, Fred. 2015. *Big History and the Future of Humanity, Second Edition.* Oxford, Wiley-Blackwell.

Spier, Fred. 2019. ‘Chapter one: Humanity has a choice: Our common future from a big history perspective.’ In: Radia, Pavlina, Winters, Sarah Fiona, & Kruk, Laurie (eds.) *The Future of Humanity: Revisioning the Human in the Posthuman Age.* London, Rowman & Littlefield (15–27).

Thomas, Julia Adeney, Williams, Mark, & Zalasiewicz, Jan. 2020. *The Anthropocene: A Multidisciplinary Approach.* Cambridge, UK, Polity Press.

Vries, Bert de & Goudsblom, Johan. 2002. *Mappae Mundi: Humans and Their Habitats in a Long-Term Socio-Ecological Perspective, Myths, Maps and Models.* Amsterdam, Amsterdam University Press.

Wang, Jida, et al. 2018 ‘Recent global decline in endorheic basin water storages.’ *Nature Geoscience* 11 (926–932). DOI: 10.1038/s41561-018-0265-7.

Williams, Mark, Zalasiewicz, Jan, Haff, P.K., Schwägerl, Christian, Barnosky, Anthony D., & Ellis, Erle. 2015. ‘The anthropocene biosphere’ *The Anthropocene Review* 2, 3 (196–219).

Zalasiewicz, Jan. 2008. *The Earth After Us: What Legacy Will Humans Leave in the Rocks?* Oxford, Oxford University Press.

Epilogue

ONE MORE DEMOTION OF OUR OWN SPECIES?

In the *Introduction* to my *Sketch of Study Plans* (in Dutch) written in September of 1984, which outlined the preliminary plans for my cultural-anthropological research in Peru, I wrote the following sentences:

> All of life on Earth is interconnected – and is also connected to 'dead' nature. Humans are inextricably linked to the terrestrial ecological system. This system is neither static nor completely stable. After the Earth had come into existence there was a time without any life-forms. Step-by-step, a great variety of living species emerged, competing and collaborating, succeeding one another, reacting to changes in climate, while causing climate change, etc.
>
> Measured on the geological timescale, the human species has only existed for a short time. Yet during that short period we have changed the face of the Earth faster than any other species before us. Exactly in our lifetime these changes are accelerating. We are living with a growing number of people in a world with a limited amount of resources. The industrialized countries sometimes possess such resources themselves, but they are also able to channel many of those resources to themselves from somewhere else.
>
> Our industrial way of life is leading to an increasing exhaustion of the natural resources. In addition, we are producing an inevitable increasing pollution of the environment. I think that in the coming years these problematic issues will be felt more strongly all around the world. There are only few countries with abundant natural resources, while those are exploited almost everywhere at a large scale. It seems to me that the human species is very busy, and at a growing rate, to diminish the possibilities for a continued existence in reasonable well-being.
>
> To gain a better understanding in these matters, it is necessary, in my opinion, to study how people think about nature; how they are dealing with it; as well as which social interdependencies contribute to the developments mentioned.

Those 1984 reflections, still written on a typewriter, summarized my motivation for going to Andean Peru: to find out how people lived who were more directly dependent on their natural environment because they were mostly self-supporting farmers who grew most of crops that they were eating. Would such people have different attitudes toward their natural environment than urban people, including being more careful with it, because if those dependencies deteriorated, they would experience that immediately?

That is what motivated me to go to Peru to become acquainted with Andean farmers, their way of life, and their history (Spier 1994, 1995). Little did I suspect in 1984 that almost forty years later, driven by the same motivation, I would write a history of the biosphere. My ecological concerns started after seeing for the first time, in January of 1969, a photo of Earth at a distance, namely the famous Apollo 8 Earthrise photo (Spier 2019a). Those concerns led to my Peru studies as well as to my subsequent involvement in big history, starting from 1993. And this current book came as an unplanned result of both projects. It was conceived while growing peppers in Amsterdam in 2017 and observing what they were doing.

Over the past forty years, I have encountered many instances in which people, including scholars, appear to think, perhaps unconsciously, that our modern human lives are, to some extent, separate from the biosphere. That the biosphere is 'out there,' to some extent removed from our daily urban experiences, and that we would need to preserve its integrity (whatever that may mean, seen from the perspective of this book). This bias will be called Bias # 11. If anything, I hope that this book shows that we humans are an inextricable part of the biosphere.

Could, as a result, the theoretical approach proposed in this book perhaps signal the next 'demotion' of our perception of humanity's place within the cosmos? The term 'demotion' used in such a way was introduced by the US author Ann Druyan (1949–), the late Carl Sagan's spouse.[1] It expresses the idea that as a result of scientific exploration, over the course of time humans have lost their perceived privileged positions within the cosmos: first as being at the center of the universe; next, being at the center of our solar system; and subsequently, their exceptional position within the realm of life, as described by the biblical account of creation. By looking at our biosphere's history in the way presented in this book, have humans now become a temporary important, but in all likelihood evanescent, aspect of the biosphere's history?

A FEW FINAL THEORETICAL CONSIDERATIONS

While writing this book, I came across a few theoretical issues for which there did not appear to be a place within the main text. Yet they seem sufficiently important to mention them at the end of the book. First of all, the emergence of complex cif loops during the biosphere's history led to densely and intensively interacting webs of such loops, much like how the great many biochemical processes interact within cells. In the biosphere's history, this resulting dynamic is clearly an important issue, which should, as a result, receive much more attention in the future.

Yet while contemplating this issue, I decided that it was currently beyond my grasp to be more precise about it, not least because the analysis of the complex cif loops itself is still very preliminary and rudimentary. For instance, how would we determine the complexity of all those loops as well as the characteristics of their resulting joint dynamics, especially perhaps, but certainly not exclusively, during human history?

During the earlier long history of the biosphere, it could be stated that those complex cif loops as well as their resulting webs and dynamics were becoming more diverse and complex over the course of time. But was that also the case during human history? In some ways, those webs of complex cif loops probably became more diverse and complex. But what happened to the webs and their dynamics that were, for instance, related to the decline of the undomesticated food-capturing pyramid? I found that hard, if not impossible to state with any precision. That is why I decided to leave this important subject out of my analysis, while leaving it to the next generations to tackle such issues.

[1] Carl Sagan, video of his lecture 'The Age of Exploration' https://www.youtube.com/watch?v=6_-jty-hAVT In that video, Carl Sagan credited his wife Ann Druyan for using the concept of 'demotion' with such a meaning.

A second theoretical issue emerged while posing the question how many nonlinear processes, complex cif cycles and their associated webs could be recognized in all of big history, in other words: far beyond the biosphere's history, starting with the origin of the universe. Examining this issue looks like another major challenge that could become the subject of another book (which I am probably not going to write).

Another theoretical issue arose while contemplating the process of zooming in and out within academia: looking at details at one moment while at other moments seeking to obtain overviews that accommodate all those details. During the late eighteenth and the first half of the nineteenth century, such academic attitudes were common especially among well-educated traveling naturalists such as Georg Forster, Alexander von Humboldt, and Charles Darwin, while also captain Robert Fitz-Roy possessed such an attitude. The earlier Portuguese and Spanish cosmographers also fostered them, most notably perhaps the famous Spanish naturalist, historian, and Jesuit priest José de Acosta (1540–1600), while other European cosmographers who followed in their footsteps had acquired a similar outlook.

Those wide-ranging attitudes, combining overviews with details, almost disappeared in the second half of the nineteenth century as a result of the growing academic specialization. For instance, while I was studying biochemistry in the 1970s, and cultural anthropology in the 1980s, such wide-ranging studies were considered almost forbidden territory, while zooming in and out was only allowed within what was defined as their own disciplines.

As part of that, I still vividly remember that specialists investigating half of the DNA of a bacterial virus called Lambda were not acquainted with the work of colleagues who were studying the other portion (at the time known as 'the left and right hand of Lambda'). That is how far specialization had proceeded within the field of biochemistry in those days. But at least all those biochemists shared the same general theoretical approaches.

And when I studied cultural anthropology in the 1980s, while asking about the importance of geographic influences for how people made a living, and, in consequence, their cultures, I was told by a teacher that this subject was not part of the discipline of cultural anthropology. If I were interested in such aspects, I would need to consult another discipline, namely social geography. And in this case, there were no shared general theoretical approaches between those two disciplines. To be sure, not all practitioners of the natural and social sciences were like that. But it was the dominant academic attitude in those days, at least in the places where I lived.

Today, with the emergence of big history as well as other large-scale approaches, attempts are made to recover the older more broad-minded attitudes, in doing so seeking novel forms of how to synthesize our academic knowledge. This book should be seen as part of that trend. For promoting such views, the academic world could profit a great deal by considering images of Earth at a distance. Yet for many academics, even today such broader approaches are still considered forbidden territory, especially perhaps within the social sciences and the humanities.

There appears to be a general principle that stimulates such broader mental attitudes, perhaps first outlined as such by Alexander von Humboldt in his book *Cosmos*, volume 2 (1849). The great Prussian naturalist saw it as follows. While the known world expanded through travel, conquest, trade, and migration, so did people's worldviews, at least in the perceptions of those people who wanted to obtain such overviews. This also happened during the period of early globalization, most notably as a result of the discovery of the Americas, as seen from a European perspective, and, more recently, with spaceflight that produced the first images of our home planet at a distance.

To avoid any possible misunderstandings, I would like to emphasize that in this book I did not seek to dethrone Charles Darwin's magnificent studies. To the contrary, he needs to stay on the throne where he rightfully belongs. But also Vladimir Vernadsky deserves a similar global throne. Combining their insights while adding a few of my own may have led to the beginning of a synthesis of their views within an overarching theoretical approach.

While writing this book, I encountered a great many issues and questions that for lack of space needed to remain unattended. For instance, my early doubts about the city of Eindhoven's history, mentioned in Chapter 1 and further explored in Appendix 1, led to a reappraisal of many, if not all, forms of history, including their uses, goals, and effects. I keep reflecting a great deal on all of that, most notably perhaps the prestige aspects of history, while it remains remarkable that humans appear to have been the only species to have lived on this planet with a conscious awareness of history that goes beyond personal experiences.

That is a major aspect of human life, which considerably contributes to making it so extraordinarily complex. Perceptions of history, and of reality in general, are extremely important for the definition of one's own situation; of one's own identity as well as those of others; as well as one's goals, including whether and how to act. It is therefore not surprising that views of reality and its past are highly sensitive issues for most, if not all people, and that efforts to control or contest them have existed as long as we know.

It has been a long road after starting to grow pepper plants in the spring of 2017, while observing them and contemplating what they were doing. In retrospect, I engaged in that activity also in an attempt to remain, at least in my own perception, part of the Andean society that I had become so intimately acquainted with, which has influenced me personally to the deepest possible extent, even though, by necessity, for many years I have been living far away from them.

As a final reflection: however insignificant my current contribution may be compared to the earlier studies of our far more illustrious predecessors, I very much hope that this book will assist other scholars in pursuing these fields of enquiry further and deeper, and that, in doing so, they will improve our knowledge about the biosphere's history, life, and humanity, which may hopefully contribute to the longer-term survival of humanity in reasonable well-being.

BIBLIOGRAPHY

Acosta, José de. 1987. *Historia natural y moral de las Indias (ed. de José Alcina Franch)*, Crónicas de América 34. Madrid, Historia 16 (1590).

Humboldt, Alexander von. 1849. *Cosmos: Sketch of a Physical Description of the Universe*, Volume II. London, Longman, Brown, Green, and Longmans, John Murray (original German edition: 1847).

Poole, Robert. 2008. *Earthrise: How Man First Saw the Earth*. New Haven, CT & London, Yale University Press.

Spier, Fred. 1994. *Religious Regimes in Peru: Religion and State Development in a Long-Term Perspective and the Effects in the Andean Village of Zurite.* Amsterdam, Amsterdam University Press. A pdf of this book can be downloaded for free at: http://www.bighistory.info/bhi_005_035.htm.

Spier, Fred. 1995. *San Nicolás de Zurite: Religion and Daily Life of an Andean Village in a Changing World.* Amsterdam, VU University Press. A pdf of this book can be downloaded for free at: http://www.bighistory.info/bhi_005_035.htm.

Spier, Fred. 2010. *Big History and the Future of Humanity.* Oxford, Wiley-Blackwell.

Spier, Fred. 2015. *Big History and the Future of Humanity, Second Edition.* Oxford, Wiley-Blackwell.

Spier, Fred. 2019a. 'On the social impact of the Apollo 8 Earthrise photo, or the lack of it?' *Journal of Big History* 3, 3 (157–189). https://jbh.journals.villanova.edu/article/view/2425.

Spier, Fred. 2019b. 'Chapter One: Humanity has a choice: Our common future from a big history perspective.' In: Pavlina Radia, Sarah Fiona Winters, and Laurie Kruk (eds.) *The Future of Humanity: Revisioning the Human in the Posthuman Age.* London, Rowman & Littlefield (15–27).

Appendix 1
A Personal Look at the City of Eindhoven's History

Based on my personal observations in Eindhoven during the 1950s and early 1960s; on internet research in 2019; on the literature mentioned below; on period photos and home movies in our possession; and on my reflections and interpretations, based on all of that, Eindhoven's history can be summarized as follows.

This city is situated in the southern Dutch province of North Brabant, at the confluence of two rivulets, the Dommel and the Gender, that both provided water for its inhabitants, for their cattle, for their fortifications, as well as for their proto-industrial needs. This includes the Genneper watermill along the Dommel rivulet painted by Vincent van Gogh in 1884. After many centuries, this watermill had ceased operating when, in 1955, our little family settled nearby in a modern and relatively wealthy residential area which was part of that Gestel neighborhood.

By 2019, this mill had nicely been restored as a museum, with a functioning water wheel. But in the 1950s, that abandoned mill looked to me a dreary and gloomy place straight out of an unpleasant fairytale. Yet dark and dirty as it was, it also fascinated

The Genneper watermill in 1940. (Postcard by REB, in possession of the author)

me as a remnant of the area's dark past that was unknown to me. Furthermore, the water mill featured a type of technology that I could immediately understand, while I could easily imagine how its commercial operations might have worked. All of that was the beginning of my life-long interest in both history and technology.

To be sure, watermills are free-energy capturing devices. The rather limited free-energy gradient provided by the Dommel rivulet flowing within an almost flat landscape had, for centuries, been considered sufficiently valuable for tapping it, while using it for productive purposes. But the industrial revolution had provided far more abundant and cheaper free-energy flows, which led to the abandonment of the Genneper watermill.

This was all part of Eindhoven's deeper history. As early as 1232, this little town had acquired city rights, which was earlier, in fact, than most Dutch cities in the western part of the country which later became dominant. However, Eindhoven's surrounding land was not very fertile. It mostly consisted of sandy soils surrounded by swamps, while there were no other major natural resources either. As a result, the local elites could not accumulate great wealth from the local and regional resources, while the relatively cheap labor force consisted of mostly poor people, many of whom were probably often looking for new ways to make a living.

The little city of Eindhoven was also situated on the intersection of two long-distance roads: one running from the north to the south and the other one from the west to the east. This provided a flow of travelers, including merchants, many of whom must also have carried information about what happened elsewhere. Like so often in human history, by serving the travelers' needs this intersection turned into a little commercial hub.

As part of that development, small artisan shops as well as a few proto-industrial enterprises emerged, such as leather workers, blacksmiths, rope and cloth producers, bakeries and breweries, butchers, and later also snuff production mostly for elites to the south. And especially from the beginning of the industrial revolution, many

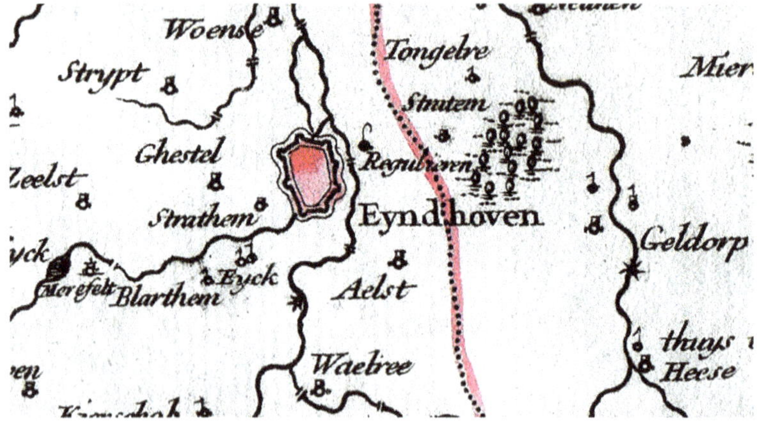

Fragment of a map of the Eindhoven area, the city depicted with walls and a moat. (From: Carte du Brabant ou se trouv.[t] les envir.[s] d.'Eyndhoven dans la Mairie de Bosleduc. Par le S.[r] Robert Geographe ord. Du Roi avec Privilege 1748, original in possession of the author)

cigar factories were set up in the area, making use of the cheap and abundant labor, while also the production of hats took off, including considerable numbers of silk top hats by the Spoorenberg firm.[1]

In the 1950s, their main hat store was located at the Stratumseind street, where the main road to the south left town. This was just around the corner of the Philips Tungsten and Molybdenum Factory where my father then worked. Because their earlier facility had been bombed out during the Second World War, this Philips factory was relocated to that building, which had earlier housed a leather factory. There, tungsten and molybdenum were purified, both metals with a high melting point that were used for making the filaments of incandescent lamps, which had been Philips's original core business. It was my father's task as a chemical engineer to help improve the strength and durability of those filaments (cf. H.L. Spier 1961).

The Stratumseind street, much as I remember it. (Postcard by REB, postmarked June 24, 1959, in possession of the author)

[1] The video by the Spoorenberg firm from 1950: 'The making of a silk top hat': https://youtu.be/6j2Mbz7WVwI offers a remarkable example of a proto-industrial mode of production in Eindhoven, then operating in the shadow of a large and rapidly-growing globally-oriented electronics firm.

That Philips factory was located in the Molenstraat, 'Mill street,' named after the nearby Stratum watermill in the Dommel rivulet that had existed there at least since 1340, while it had been demolished as recently as 1928. So, by the time my father started working there in 1945, the Molenstraat area had already known a history of at least 600 years of proto-industrial activities. This was unknown to me until the fall of 2019, when I finally began looking into Eindhoven's deeper history.

In 1955, my mother and us three little boys used to walk along the Dommel rivulet to that Philips factory to pick up my father in the evening. I remember that production area as one big hell hole full of black soot and burning fires, and the only remaining photo of that scene in my possession certainly confirms those impressions. By contrast, the office area where my father worked looked reasonably clean and OK to me.

A few years later, the factory moved to a newly-built facility near the village of Maarheeze, situated some 25 km to the southeast of Eindhoven. Although much of that then new and iconic building no longer exists, the former Philips lighting division, now called Signify, is still in business there as a major player on the world stage.

The production facility of the Tungsten & Molybdenum Lab in the Molenstraat, 1950s. (From the commemorative booklet *Ter herinnering aan het Wolfraam en Molybdeen Lab, 1945–1963*, produced for my father's farewell from Philips in 1963, original in possession of the author)

The office of the Tungsten & Molybdenum Lab in the Molenstraat, 1950s. Up front, Mr. Vergeer, on the left, my father Henri Louis Spier. (From the commemorative booklet *Ter herinnering aan het Wolfraam en Molybdeen Lab, 1945–1963*, produced for my father's farewell from Philips in 1963, original in possession of the author)

In the 1950s, my father used to buy his Borsalino hats at the Spoorenberg hat store. I remember going there, accompanying him, and watching all of that with a certain puzzlement, because to me those hats looked ancient and prehistoric, and I could not imagine myself ever wearing such a hat even when I became that old. Yet at that time, those hats were actually rather trendy and fashionable among certain social classes.

Returning to Eindhoven's deeper past: because of its geographical location as well as the yields of its relatively poor agriculture and proto-industrial activities, the little city soon became attractive to potential conquerors. But its limited resources did not allow such warriors to keep that city for long. As a result, over the centuries the city changed hands many times, sometimes within a few years, most notably perhaps during the northwestern Dutch uprising against the Spanish Habsburg house (the so-called "Eighty Years' War" between 1568 and 1648). All those changes beyond their control must have had effects on the local population's mentality.

After in 1648 the area had come under formal control of the Dutch Republic to the north as part of the Peace of Westphalia, while its city walls and little fortress had

My father, my older brother Jaap, and me c.1955. (From: home movie X1b, in possession of the author)

been destroyed, Eindhoven's modest urban aspects declined, even though its proto-industries continued. That situation may have led to forgetting about Eindhoven's earlier city past.

During the nineteenth century, the industrial revolution began to have an increasing impact on that area, not least because it provided new means of production, transportation, and communication, all based on the far greater availability of free energy as well as its lower price. As a result, many people became less dependent on local and regional resources, while they began to rely more on global interdependencies.

After first a canal, and, a little later, also a railroad had reached Eindhoven, while the industrial uses of electricity were taking off worldwide, in 1891 the wealthy banking and trading Philips family from the small town of Zaltbommel to the north started the production of incandescent lamps in Eindhoven, in doing so making use of its proto-industrial past, its cheap labor force, as well as its recent industrialization by other firms, most notably perhaps factories that made cigars and matches. In fact, up until 1910, cigar factories remained the most prominent type of industry in Eindhoven.

All those developments were part of a trend in which entrepreneurs from elsewhere kick-started the industrialization of that area, while local manufacturers such as the Spoorenberg hatters stuck to more traditional modes of production. After the Second World War, also the DAF truck company began to flourish nearby, which was another effect of Eindhoven's favorable location on that road intersection as well as of the nearby presence of the Philips company and its growing transportation needs.

Yet during the first half of the twentieth century, it was, most notably, the rapid change of the Philips electronics company from a little local enterprise into a global electronics powerhouse that suddenly transformed the city of Eindhoven into a prominent industrial hub with worldwide aspirations. Its earlier and more modest industrial enterprises were shut down step-by-step and were soon largely forgotten, while their

The Philips 'Light Tower' in Eindhoven in the 1950s, then the home office of the rapidly-growing global electronics firm. (Postcard Uitgeverij Takken, Utrecht, in possession of the author)

material heritage was almost entirely erased by city planners who sought to transform Eindhoven into a modern industrial city. That was the situation when we lived there between 1955 and 1963 (for me, between three and ten years of age).

It may well be that during those years, only a few Eindhoven inhabitants had an overview of that city's history such as the one sketched here. Surely, none of that was taught at my primary school, the (non-religious) Nutsschool in the Akkerstraat situated in the Gestel neighborhood, a few streets of which had turned into a virtual enclave of Philips engineers and their families.

The dominant story of Eindhoven's history told then (as I remember it) was that the city had recently grown from the five rural hamlets (including Gestel) that surrounded the equally little and insignificant rural village of Eindhoven, all of which had recently been merged into the current city as a result of Philips' expansion. To some extent, this version of history is still displayed on the City of Eindhoven's website, although it also pays some attention to its earlier past. In other words, it was the history of Eindhoven after cheap and abundant free energy had become available by the end of the nineteenth century which led to the emergence of industrial production.

In the 1950s, telling that rather limited version of Eindhoven's history went hand-in-hand with Philips' expansion and the rapid introduction of a great many novel industrial products that dramatically changed people's daily lives, ranging from washing machines and refrigerators to radios and television sets, while they also included mopeds and cars, gramophones, and tape recorders, a great many medical applications, as well as many forms of automation, while there were also ever more uses found for synthetic materials such as plastics.

For many people, those enormous technological and social changes produced a clearly-felt break with the past, while they stimulated a forward-looking mentality with great technological and social expectations for the future, in which 'progress' and 'modernity' were key concepts. Such an outlook was then common among people around the world who experienced such developments, while the past was increasingly seen as consisting of episodes of an earlier period that was now being left behind forever.

As part of that, most of Eindhoven's earlier material heritage was erased and replaced by what was considered more appropriate modern architecture. All of this has made it hard, if not almost impossible, to 'read' the older history of that city by walking through it, in ways in which this still can be done in a great many other traditional cities both in Europe and elsewhere. Only if one knows what to look for in Eindhoven, such as street names and their patterns, can some of those clues be recognized, while a walk along the Dommel rivulet will provide an impression of remnants of its earlier industries.

Modernity in Eindhoven c.1963: The prestigious Hotel Cocagne, and a DAF car with a novel automatic 'smart gearshift.' (Photo by my father H.L. Spier, original negative scanned by and in possession of the author)

Seen from a sociological point of view, as a result of its industrial growth in the twentieth century, Eindhoven's population came to consist of a large and powerful group of industrial newcomers with diverse cultural orientations, who increasingly put their stamp on the city, including their version of its history, while a considerable portion of the older-established and mostly Roman Catholic population was marginalized, even though their numbers were also growing.[2]

Those people may have reacted, among other things, with reviving Catholic processions and other 'traditional' celebrations such as carnival and 'banner swinging' parades ('vendelzwaaien' in Dutch), also into the newer residential areas where the Catholic presence was limited, in doing so claiming those areas symbolically, as it were. One such a parade was filmed in the summer of 1955 by my father from the safety of the first floor of our house, which offered a good overview. We still have that movie, now digitized. I am not sure whether any of my family then recognized that they were watching such a process.

I remember more such events, including the one that I attended at six or seven years of age (or so I think), when I spontaneously decided to join in a passing religious procession toward the nearby newly-built Catholic secondary school, the St. Catharina Lyceum, named after St. Catharine of Alexandria, who was already an important patron saint in the city. I was attracted, but also bewildered, by the religious paraphernalia that those people were carrying, without having any idea what they meant or what they were used for, other than that apparently, those people were Catholics, and that all of that probably had to do in some ways with God and Jesus.

Banner swinging in Eindhoven in the summer of 1955, filmed by my father from the first floor of our house. (From: home movie X1-a, in possession of the author)

[2] At that time, in Eindhoven there was also a sizable Protestant population, to a considerable extent an effect of the dominance by the Dutch republic. Yet the majority of the local population was Catholic.

At a certain moment, on the square in front of that college, everybody started kneeling while facing a baldachin, underneath of which artful things were shown made out of silver and gold (now I think that this probably included a monstrance displaying the host). The participants strongly urged me to kneel as well, which I did, without having any idea why that would be necessary or what was going on, other than that it was apparently some sort of holy veneration that I did not know and even less understood. Surely, those people felt like complete strangers to me, as if I were suddenly participating in a ritual among visitors from another planet where I did not belong. After that experience, I never joined a Catholic procession again in Eindhoven.[3]

Yet the city center's skyline was – and to some extent still is – dominated by a number of large Catholic churches, while there were also many other Catholic institutions such as schools and convents all over the city and beyond, most of them dating back to the late nineteenth and the first half of the twentieth century, when Eindhoven's Catholic population was rapidly growing. Those ecclesiastical structures could only be built after the Dutch constitutional reform of 1848 had allowed Roman Catholicism to become a public religion again. During the mid-twentieth century, when we lived in Eindhoven, those buildings were no-go areas for the growing numbers of Philips employees who were not Catholics, many of whom (like my family) had migrated to the city from other Dutch regions.

As a result, many Philips employees may have developed a dual identity at that time: a strong Philips identity, and a weaker Eindhoven one. For me, that was certainly the case. I acquired a rather strong identity with the remarkably tightly-organized community of modest, friendly, and competent Philips engineers who embodied technical and social progress, among whom religious and political orientations did not play any detectable role.

Yet my identity with the city of Eindhoven was far less pronounced and clear because those Catholic people appeared to belong to another world of which we knew very little. As a result, we rarely, if ever, went to the Eindhoven areas where those people lived – where there were still remnants of the city's deep past – unless we needed to go there for practical reasons such as shopping. Yet they always treated us well in their typically Brabantian modest and friendly ways.

Wearing hats like my father did may also have been a – perhaps unconscious – expression of belonging to the new powerful and prestigious elite, while many of the poorer laborers wore nothing on their heads, or sometimes caps. A photo in my possession of the Philips Tungsten and Molybdenum factory personnel taken around 1950 clearly shows those differences.[4]

---

3 Since that time, I have come a long way in reaching a deeper understanding of the world of Roman Catholicism, first thanks to Mart Bax's enlightening research into religion and politics in North Brabant in the 1980s, and subsequently through my own research into a similar theme in Andean Peru.

4 Such social differences expressed in wearing hats or caps was a common situation during the industrial revolution in Europe and elsewhere. This also raises the question of the sudden disappearance of wearing such hats starting from the end of the 1960s and the subsequent regentrification of wearing caps, at least to some extent, a process that is still going on today. I call it regentrification, because around 1900, fashionable caps were also worn by higher-middle class young adults during excursions, a habit which might mostly have disappeared by the 1950s. The changing shapes of cars, more streamlined and thus with lower roofs and consequently with less head space, may have played a role in this regentrification process, because this left less room for wearing hats.

The Lange Haven (Long Harbor) in Schiedam, c.1960s, its epicenter of gin distilling for centuries. (Postcard Hema, in possession of the author)

In 1963, my parents moved to the city of Schiedam, situated in the Western Netherlands, while my father went to work at the Unilever Research Laboratory in the neighboring city of Vlaardingen. In Schiedam, my Eindhoven past (including Philips) did not matter anything at all to the local population, presumably also because it was not part of the Dutch national history school curriculum. Furthermore, the behavior of many of my Schiedam peers was considerably rougher and self-assertive compared to what I had experienced in Eindhoven. All of that caused a culture shock in me, because I did not know how to handle that situation satisfactorily.

To survive those circumstances, I repressed most of my Philips and Eindhoven identities while seeking to adapt to that new situation. Yet regardless of my efforts, I remained an outsider among most of the Schiedam elite, even though virtually all of them lived in the same neighborhood. Some of those families had resided in Schiedam for centuries while making a living by producing a strong liquor called *jenever* (gin) as well as by ship building and maintenance, while others earned their money by serving in the legal and medical professions. For me, that situation produced only a limited positive identity with the city of Schiedam.

BIBLIOGRAPHY

Bekooy, Guus, Derks, Sergio, Haneveer, Jeff S. & van der Put, Frans (eds.). 1991. *Philips Honderd, 1891–1991.* Zaltbommel, Uitgeverij Europese Bibliotheek b.v.

City of Eindhoven. 2019. *De Geschiedenis van Eindhoven.* https://www.eindhoven.nl/stad-en-wonen/stad/erfgoed/de-geschiedenis-van-eindhoven (in Dutch).

Elias, Norbert & Scotson, John L. 1994. *The Established and the Outsiders: A Sociological Enquiry into Community Problems.* London, Sage Publications (1965).

Govers, Jacques & van der Sommen, Willem. 2000. *Eindhoven in oude ansichten - een wandeling door het centrum*. Zaltbommel, Europese Bibliotheek.

Govers, Jacques & van der Sommen, Willem. 2002. *Eindhoven in oude ansichten - een wandeling door Gestel*. Zaltbommel, Europese Bibliotheek.

Oorschot, dr. J.M.P. van. 1993. *Het Gezicht van Nederland: Eindhoven*. Abcoude, Uitgeverij Uniepers. (By far the best summary of Eindhoven's history known to me).

Spier, Henri Louis. 1961. *Influence of Chemical Additions on the Reduction of Tungsten Oxides*. Ph.D. Thesis, Technische Hogeschool Eindhoven. The full text can be downloaded at: https://pure.tue.nl/ws/portalfiles/portal/3576744/73050.pdf.

Vermeeren, Karel. 1977. *Eindhoven: Toen Eindhoven nog Eindhoven was*. Den Haag, Kruseman's Uitgeversmaatschappij B.V.

Appendix 2
List of Biases Mentioned in the Book

Bias # 1: Many urban people may think that the 'goal' of plants in the wild is to produce fruits and seeds, while in reality their major 'goal' is survival (Chapter 1).

Bias # 2: Most biologists consider the theory of natural selection to be the central theory of the history of life, which it is not. It is a theory about the origin of species (Chapter 1).

Bias # 3: To consider eating to maintain the complexity of living bodies as a major goal without sufficiently considering the most important aspect of capturing free energy and matter in shaping those bodies (Chapter 3).

Bias # 4: The tendency to pay attention to more complex species instead of examining the entire food-capturing web (Chapter 3).

Bias # 5: The preference for examining the individuals of certain species, and certain species as a whole, while neglecting to investigate them systematically as part of the network of interdependencies that connects them to all other species (Chapter 3).

Bias # 6: Equating the 'real' beginnings of processes with the period of which we have the earliest-available evidence of them (Chapter 4).

Bias # 7: Thinking that the major trend in biological evolution would be the emergence and development of increasingly-complex life-forms, culminating with humans (Chapter 4).

Bias # 8: The tendency to pay attention to Cambrian animals, the secondary and tertiary captors of free energy, while neglecting the Cambrian plants, the primary captors of free energy (Chapter 5).

Bias # 9: The focus in human history on individuals and individual 'societies' that are descended from each other, while all the rest is considered the 'environment,' if at all, instead of regarding all human societies as well as their 'natural' environment as part of one single greater whole, the biosphere (Chapter 6).

Bias # 10: To focus the attention in human history on the greater complexity of 'civilizations,' most notably their heroic deeds and cultural achievements, while neglecting all the other people, most notably the farmers, slaves, and other workers, even though until very recently such people usually made up about 90% of such entire populations, while providing most of the free energy and matter that has made the human food-capturing pyramid possible (Chapter 6).

Bias # 11: The often unspoken idea that the biosphere is 'out there,' and that our human lives and all their effects are not an integral part of it (Epilogue).

Index

Note: **Bold** page numbers refer to tables; *italic* page numbers refer to figures and page numbers followed by "n" denote footnotes.

For Product Safety Concerns and Information please contact our EU representative GPSR@taylorandfrancis.com
Taylor & Francis Verlag GmbH, Kaufingerstraße 24, 80331 München, Germany

www.ingramcontent.com/pod-product-compliance
Lightning Source LLC
Chambersburg PA
CBHW060349220326
41598CB00023B/2860

* 9 7 8 1 0 3 2 2 3 0 4 0 5 *